# Cables
# and Wiring

# Cables
# and Wiring

**AVO Multi-Amp® Institute
and John Cadick**

Delmar Publishers Inc.

## NOTICE TO THE READER

Publisher does not warrant or guarantee any of the products described herein or perform any independent analysis in connection with any of the product information contained herein. Publisher does not assume, and expressly disclaims, any obligation to obtain and include information other than that provided to it by the manufacturer.

The reader is expressly warned to consider and adopt all safety precautions that might be indicated by the activities described herein and to avoid all potential hazards. By following the instructions contained herein, the reader willingly assumes all risks in connection with such instructions.

The publisher makes no representations or warranties of any kind, including but not limited to, the warranties of fitness for particular purpose or merchantability, nor are any such representations implied with repect to the material set forth herein, and the publisher takes no responsibility with respect to such material. The publisher shall not be liable for any special, consequential or exemplary damages resulting, in whole or in part, from the readers' use of, or reliance upon, this material.

Cover design by design M design W
Cover photo courtesy of Rome Cable Corporation

**DELMAR STAFF**
New Product Acquisitions: Mark W. Huth
Assistant Editor: Nancy Belser
Project Editor: Cynthia Lassonde
Production Supervisor: Wendy Troeger
Art Coordinator: Brian Yacur
Senior Design Supervisor: Susan C. Mathews

For information, address Delmar Publishers Inc.
3 Columbia Circle, Box 15-015
Albany, New York 12212-5015

**Delmar Publishers' Online Services**
To access Delmar on the World Wide Web, point your browser to:
**http://www.delmar.com/delmar.html**
To access through Gopher: gopher://gopher.delmar.com
(Delmar Online is part of "thomson.com", an Internet site with information on more than 30 publishers of the International Thomson Publishing organization.)
For information on our products and services:
email: info@delmar.com
or call 800-347-7707

5   6   7   8   9   10   XXX   99   98   97   96

Printed in the United States of America
Published simultaneously in Canada by Nelson Canada,
a division of The Thomson Corporation

**Library of Congress Cataloging-in-Publication Data**
        p. cm.
    Includes index.
    ISBN 0-8273-5460-6
    1. Electric wiring. 2. Electric cables. 3. Electric wire.
I. AVO Multi-Amp Institute.
TK3201.C33   1993
621.319'24—dc20                    92-31886
                                        CIP

# Table of Contents

## PART ONE

Covers general wire and cable theory, as well as cable materials and construction, installation, splicing, termination, and testing procedures.

*Chapter 1:*

*Chapter 2:*

## PART TWO

Includes specific cable application information, such as cable sizes, current capacities, permitted uses, and applications.

*Chapter 6:*

# Preface

*Cables and Wiring* was written as a reference book for journeymen and apprentice electricians. This book provides fingertip access to power cable and wire construction, splicing, termination, testing, and applications.

Written in two parts, *Cables and Wiring* begins with chapters that illustrate, in detail, acceptable methods for installing, splicing, and terminating power cables.

The second part of *Cables and Wiring* is a complete reference to over 25 types of Underwriters Laboratories® and National Electrical Code® listed cable and wire. Information provided includes: available sizes, ampacities, temperature ranges, allowable installations, receiving and handling, and termination methods.

Wire is the most common type of electrical equipment. When electricity is compared to water, wires are the pipes that pass the "electrical fluid" from where it is to where we want it to be. The electrical conductor compares to the opening in the pipe and the insulation is similar to the wall of the pipe.

While the principles of wire are simple in concept, diverse applications and service requirements have led to a wide variety of wire types, sizes, shapes, and structures. In addition to the different sizes for various loads, industry has also developed wire for wet environments, wire for corrosive environments, wire for hot environments, and wire for cold environments. Each of the different types of wire and cable have different handling, installation, and testing requirements.

All of this complexity led to confusion and misunderstanding. The problem was aggravated by the fact that there were no good, concise references available for the practicing electrician, until now!

Intended for the practicing journeyman and apprentice electrician, *Cables and Wiring* provides information that is unavailable in any other single location. Starting with a simple, yet comprehensive explanation of the fundamentals, the book advances to chapters detailing cable materials and construction, installation, testing and fault location, and a detailed synopsis of over twenty-five *National Electrical Code*® (*NEC*®) wire and cable types. Sections are included on appplication, sizes, temperature ranges, handling and care, installation, splicing and terminations, and testing for each of the major *NEC*® cable types.

When the electrician has questions about a particular type of wire and cable, he or she need only look as far as the information chapters at the end of this handbook. *Cables and Wiring* is destined to become the electrician's guide for wiring, its types, use, applications, terminations, and testing.

# Foreword

*Cables and Wiring* was written as a joint effort between AVO Multi-Amp® Institute, the training division of AVO Multi-Amp® Services Corporation, and John Cadick, President of Cadick Professional Services.

AVO Multi-Amp® Corporation is the pre-eminent provider of high quality test equipment, testing and maintenance services, and technical training for electrical power equipment and power systems. These products and services are sold world-wide to private and government power utility companies, as well as large industrial and military facilities.

AVO Multi-Amp® began manufacturing relay and circuit breaker test equipment in 1951. AVO Multi-Amp® Corporation along with our sister companies, AVO Biddle® Instruments and AVO Megger® Instruments, Limited, are the dominant companies in the world providing power system testing, equipment, and services.

Since 1963, AVO Multi-Amp® Institute has been providing performance-based training to over 70,000 students from utilities, industry, government, and military facilities worldwide. In addition to the learning center located in Dallas, Texas, training is provided at three regional learning centers in Orlando, Philadelphia, and Toronto.

AVO Multi-Amp® Institute has also developed more than 80 technician-level maintenance training programs. All the courses were developed using the Instructional Systems Development (ISD) Method. All courses are written to target job performance requirements and are field-tested and proven.

John Cadick is the President of Cadick Professional Services, an independent electrical engineering consulting firm. Mr. Cadick is a registered professional engineer with a master's degree in engineering from Purdue University. As a practicing engineer, Mr. Cadick has been involved in design, installation, and testing of wire and cable systems for twenty-five years. He has developed and presented numerous training programs on virtually all phases of wire and cable applications.

Mr. Cadick has served as a Staff Field Engineer for Indianapolis Power and Light Company; Director of Multi-Amp® Institute; Western Regional Operations Manager for Electro-Test Inc.; and Vice-President and General Manager of Southwest Engineers of Louisiana. Since 1986, he has owned and operated Cadick Professional Services.

# Acknowledgments

AVO Multi-Amp® Institute and John Cadick, PE, wish to thank the following companies and individuals for their invaluable assistance in supplying technical information and art for this book.

| | |
|---|---|
| A. J. Pearson | National Joint Apprenticeship Training Committee |
| Jim Boyd | National Joint Apprenticeship Training Committee |
| Jack Pullizzi | AT&T Bell Laboratories |
| Sam Brown | Rome Cable Corporation |
| L. James Milne | Pyrotenax USA, Incorporated |
| S. Graeme Thomson | AVO Biddle® Instruments |
| James Marshall | 3M Electrical Products Division |
| Christopher B. Jonnes | Polywater Corporation |
| John Fee | Polywater Corporation |
| Alan Mark Franks | AVO Multi-Amp® Institute |
| Michael D. McHorse, PE | The Okonite Company |
| Priscilla West | Raychem Corporation |
| John Cavenaugh | General Cable Company |
| Allan P. Marconi | Manhattan Electric Cable Corporation |
| Richard Aure | Canoga-Perkins Corporation |
| Christine Arnold | AMP Incorporated |

A special thanks for their efforts to John J. Lapicola, President, AVO Multi-Amp® Services; Sandy Young, Educational Resources Manager; Dennis Neitzel, CPE, Senior Training Specialist; Bob Depta, Senior Training Specialist; and the entire staff of AVO Multi-Amp® Institute for their support during the development of this book.

# Dedication

This book is dedicated to Ruben Esquivel, President/CEO, AVO Multi-Amp® Corporation and John Lapicola, President, AVO Multi-Amp® Services Corporation.

# Delmar Publishers is Your Electrical Book Source!

Whether you're a beginning student or a master electrician, Delmar Publishers has the right book for you. Our complete selection of proven best-sellers and all-new titles is designed to bring you the most up-to-date, technically accurate information available.

## DC/AC THEORY

**Delmar's Standard Textbook of Electricity/ Herman**
This full-color, highly illustrated book sets the new standard with its comprehensive and up-to-date content, plus complete teaching/learning package, including: Lab-Volt and "generic" lab manuals and transparencies.
*Order # 0–8273–4934–3*

## CODE & CODE-BASED

**1993 National Electrical Code®/ NFPA**
The standard for all electrical installations, the 1993 NEC® is now available from Delmar Publishers!
*Order # 0–87765–383–6*

**Understanding the National Electrical Code®/ Holt**
This easy-to-use introduction to the NEC® helps you find your way around the NEC® and understand its very technical language. Based on the 1993 NEC®.
*Order # 0–8273–5328–6*

**Illustrated Changes in the 1993 NEC®/ O'Riley**
Illustrated explanations of Code changes and how they affect a job help you learn and apply the changes in the 1993 NEC® more effectively and efficiently!
*Order # 0–8273–5304–9*

**Interpreting the National Electrical Code®, 3E/ Surbrook**
This excellent book provides the more advanced students, journeyman electricians, and electrical inspectors with an understanding of NEC® provisions. Based on the 1993 NEC®.
*Order # 0–8273–5247–6*

**Electrical Grounding, 3E/ O'Riley**
This illustrated and easy-to-understand book will help you understand the subject of electrical grounding and Article 250 of the 1993 NEC®.
*Order # 0–8273–5248–4*

## WIRING

**Electrical Wiring—Residential, 11E/ Mullin**
This best-selling book takes you through the wiring of a residence in compliance with the 1993 NEC®. Complete with working drawings for a residence.
*Order # 0–8273–5095–3*

**Smart House Wiring/ Stauffer & Mullin**
This unique book provides you with a complete explanation of the hardware and methods involved in wiring a house in accordance with the Smart House, L.P. system. Based on the same floor plans found in Electrical Wiring—Residential, 8E.
*Order # 0–8273–5489–4*

**Electrical Wiring—Commercial, 8E/ Mullin & Smith**
Learn to apply the 1993 NEC® as you proceed step-by-step through the wiring of a commercial building. Complete with working drawings of a commercial building.
*Order # 0–8273–5093–7*

**Electrical Wiring—Industrial, 8E/ Smith & Herman**
All-new content on hazardous locations, more detailed coverage of branch circuits, and calculating circuit sizes with the 1993 NEC® will help you learn industrial wiring essentials! Complete with industrial building plans.
*Order # 0–8273–5325–1*

**Cables and Wiring/ AVO Multi-Amp®**
Your comprehensive practical guide to all types of electrical cables, this book discusses applications, storage, handling, pulling, splicing, and more!
*Order # 0–8273–5460–6*

**Raceways and Other Wiring Methods/ Loyd**
This excellent new book provides you with complete information on metallic and nonmetallic raceways and other common wiring methods used by electricians and electrical designers.
*Order # 0–8273–5493–2*

**Illustrated Electrical Calculations/ Sanders**
Your quick reference to all of the formulae and calculations electricians use, this handy book features illustrations, applications, examples, and review questions for each calculation or formula.
*Order # 0–8273–5462–2*

## MOTOR CONTROL

**Electric Motor Control, 5E/ Alerich**
The standard for almost 30 years, this best-selling textbook explains how to connect electromagnetic and electric controllers.
*Order # 0–8273–5250–6*

**Industrial Motor Control, 3E/ Herman**
This excellent third edition combines a solid explanation of theory with practical instructions and information on controlling industrial motors with magnetic and solid-state controllers. Includes coverage of programmable controllers.
*Order # 0–8273–5252–2*

## EXAM PREPARATION

**Journeyman Electrician Exam Preparation/ Loyd**
**Master Electrician Exam Preparation/ Loyd**

*To request examination copies, call or write to:*
Delmar Publishers Inc.
3 Columbia Circle
Box 15015
Albany, NY 12212-5015
Phone: 1-800-347-7707 • 1-518-464-3500 • Fax: 1-518-464-0301

# PART ONE

# Chapter 1

# General Concepts

## KEY POINTS

- What are conductors, insulators, and semiconductors?
- How does current flow through conductors, insulators, and semiconductors?
- What are the key thermal, electrical, and physical characteristics of electrical insulators?

## INTRODUCTION

All materials can be categorized based on their ability to conduct electricity.

Conductors are those materials that offer little opposition to the passage of electric current. Materials such as copper or aluminum have large numbers of "free" electrons in their structure; when a voltage is applied, the electrons move easily. The electrical resistance of such a material is a few hundredths of an ohm per one thousand feet.

In contrast, insulators have relatively few "free" electrons in their chemical structure. Materials such as rubber and plastic allow very little current flow when a voltage is impressed across them and may have resistances of several hundred million ohms per inch.

Semiconductors fall between conductors and insulators in electrical resistance. Over a limited temperature range semiconductors behave more like conductors than insulators. Although they share some of the properties of conductors, they do not conduct electricity as well. Semiconductor materials include silicon and germanium. These materials, when treated chemically, exhibit behavior that allows them to be used to make devices such as transistors and integrated circuits.

All of these materials are used in the construction of various types of wire and cable. See Chapter 2 for a more detailed description of wire and cable construction.

## TERMS

The following terms and phrases are used in this chapter. (Many of these definitions are paraphrased from *ANSI/IEEE Std 100–1988: Standard Dictionary of Electrical and Electronic Terms.*)

**Conductor**: A substance (usually metallic) that allows a current of electricity to pass continuously along it. The word may also be used to identify a wire or cable that is used in a power system to conduct electricity.

**Corona**: The luminous discharge caused by the ionization of air around a high voltage conductor.

**Dielectric Strength**: The voltage gradient at which insulation breakdown occurs and current flows.

**Equivalent Circuit**: An arrangement of circuit elements that have characteristics electrically equivalent to, but may be physically different from, those of a different circuit or device.

**Insulator**: A material that impedes or opposes the passage of electric current flow.

**Semiconductor**: An electronic conductor, with resistivity in the range between conductors and insulators.

**Surface Current**: Current that flows over the surface of an insulator.

**Tracking**: Degradation of insulation by the formation of external carbonized paths. The carbon tracks cause surface leakage current and can lead to catastrophic failure of the insulation.

**Treeing**: Degradation of insulation by the formation of internal carbonized paths. When examined in a cross sectional view, the carbonized paths resemble the branches of a tree.

**Volumetric Current**: The current that flows through insulation.

## CURRENT FLOW

### Introduction

Ohm's law is the formula that defines the basic concept of current flow. Ohm's law states that the flow of current in an electric circuit is proportional to the applied voltage and inversely proportional to the resistance. In mathematical terms Ohm's law may be shown as:

$$I = \frac{E}{R} \tag{1}$$

Where:

> $I$ = the current flow in Amperes
> $E$ = the applied voltage in Volts
> $R$ = the circuit resistance in Ohms

Although Ohm's law applies in virtually every situation, conditions may be different depending on the type of material and whether it is an insulator or a conductor. When used in electrical power applications, semiconducting material behaves very much like conducting material.

## Conductors and Semiconductors

**Equivalent Circuit.** An equivalent circuit is a diagram that is used to describe the electrical behavior of a system. The equivalent circuit for a conductor is very simple and is shown in Figure 1-1.

As the current ($I$) flows through the wire, a voltage drop ($E$) is created according to Ohm's law. The resistance ($R$) will be very low for a good conductor.

**Thermal Behavior.** The resistance of all materials varies with temperature. A conductor has many free electrons. As the temperature goes up, the electrons vibrate more actively. Because of this, the electrons literally "get in each other's way" and cause the number of collisions between them to increase. The increase in collisions translates into more opposition to the current flow and, therefore, higher resistance.

## Insulators

**Equivalent Circuit.** As you can see from Figure 1-2, the equivalent circuit for insulators is somewhat different from that of a conductor. Figure 1-2 shows the path the electrons will take as they pass *through* an insulator. The total current is composed of two components: the capacitive current ($I_C$) and the resistive current ($I_R$).

**Figure 1-1    A conductor and its equivalent circuit.**
*(Courtesy of Cadick Professional Services)*

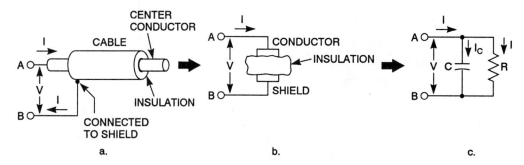

**Figure 1-2   Insulation and its equivalent circuit.**
*(Courtesy of Cadick Professional Services)*

*Capacitive Current.* A capacitor is created when two conductors are separated by a layer of insulation, the exact situation that exists in a wire or cable. Virtually all electrical power circuits exhibit some degree of capacitance; wire and cable, however, are particularly capacitive.

When a capacitor has an alternating voltage impressed on it, the resulting current leads the applied voltage by 90 degrees. The capacitive version of Ohm's law is:

$$I_C = \frac{E}{X_C} \angle 90° \qquad (2)$$

Where:

$I_C$ = the capacitive current flow

$E$ = the applied voltage

$X_C$ = the capacitive reactance

$X_c$ is the capacitive equivalent of resistance. It is a function of geometric characteristics such as the cross sectional area of the conductive plates and separation of the conductive plates.

The capacitive current is approximately 100 times greater than the resistive current in good insulation.

*Resistive Current.* The resistive portion of the current flow, through and over the surface of an insulator, follows Ohm's law as explained in equation (1). It is the result of electrons moving through and over the insulator in the same way that electrons move through a conductor.

The current that flows through the insulation is called the volumetric current. The volumetric current is caused by the motion of electrons through the insulation material. The part that flows over the surface is called the surface current. Surface current is caused by the motion of electrons through the contamination on the surface of the insulator.

Since there are far fewer free electrons in the insulator than in the conductor, the resistance of an insulator is much higher than the resistance of a conductor. The resistive current is approximately 1/100 of the capacitive current in good insulation. Resistive current in insulation is also called the leakage current.

The leakage current that flows through the insulation is called the volumetric leakage. The surface current is called the surface leakage.

*Total Current.* The total current in an insulator is equal to the vector sum of the capacitive current and the resistive current. Since the capacitive current in good insulation is so much greater than the resistive current, the total current is predominantly capacitive. Figure 1–3 is a vector diagram of the summation of the two currents.

As insulation ages, its resistive current will tend to increase; however, since the capacitive current is a function of geometry, its magnitude will remain essentially the same throughout the life of the insulation. Since good insulation has capacitive current flow that is typically 100 times its resistive current, testing the relationship between the resistive current and the capacitive current is one way that insulators can be evaluated for continued serviceability. A change in the ratio of capacitive current to resistive current indicates a problem with the insulation.

**Thermal Behavior.** As insulation is heated, its electrons become more energetic and therefore easier to displace. This results in a lowered resistance. This is one of the key electrical differences between conductors and insulators.

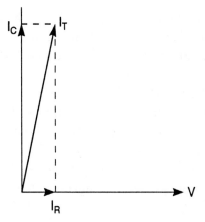

NOTE: TYPICALLY $I_C \approx 100 \times I_R$

**Figure 1–3 Vector sum of insulation capacitive current and resistive current.**
*(Courtesy of Cadick Professional Services)*

- When a conductor is heated, its resistance rises. The resistance of a conductor is proportional to temperature.
- When an insulator is heated, its resistance lowers. The resistance of an insulator is inversely proportional to temperature.

## ENVIRONMENTAL EFFECTS

### Overview

Various environmental conditions can have an adverse affect on the performance of electrical insulation. As a result, some types of cable have been developed that are resistant to these conditions. For example, silicone and Teflon® insulation have very high temperature characteristics; rubber insulation and lead sheathed cable are very resistant to moisture; and cross-linked polyethylene cable is resistant to chemical exposure. Specific information on the various cable types may be found in the reference chapters for the specific insulation and cable systems.

### Heat

When insulation is heated, its physical characteristics change, depending on the type of material that is being used. Some materials will become more flexible, while others may fail completely. If the heat is excessive, some insulation will allow the conductor to move, deforming the entire cable and degrading the insulating characteristics.

Some thermoplastic cable is also subject to a condition called treeing. Thermal and mechanical stress can cause degraded channels to form in the insulation. These channels resemble trees and allow current to flow, carbonizing them and further aggravating the problem. Eventually the cable will fail completely.

In recognition of these problems, the *National Electrical Code*® (*NEC*®)[1] places maximum operating temperature limits on insulation. Manufacturers design cable insulation with these temperature limits in mind. Modern thermosetting insulation is more resistant to heat.

Figure 1-4 shows the temperature limits of several common cable insulation systems. More detailed information can be found in the later chapters of the book, the *NEC*®, or manufacturer's literature.

---

1  National Electrical Code and NEC are registered trademarks of the National Fire Protection Association.

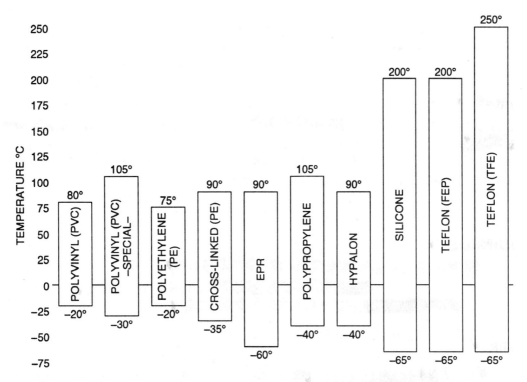

**Figure 1-4   Temperature limits of some insulations.** *(Courtesy of Manhattan Electrical Cable Corporation, Rye, New York)*

## Moisture

**Surface.** Surface moisture combined with contamination (dirt) will cause an increase in the surface leakage. These increased currents can cause two types of problems.

A phenomenon known as tracking can occur. This happens when the surface leakage currents heat the surface of the insulation causing it to turn into carbon. As the surface tracks, the currents can increase, thus aggravating the problem. Eventually, the cable can fail completely. This is a similar phenomenon to treeing that was discussed in the last section. The principal difference is that treeing occurs inside the insulation, while tracking occurs on the surface. Tracking can occur on virtually all insulation materials while treeing occurs primarily in the thermoplastic types of insulation.

Tracking and the resultant breakdown may be accompanied by corona. Corona is the luminous discharge caused by the ionization of the air around an energized conductor with a very high voltage gradient. Corona is often accompanied by the strong smell of ozone.

If moisture and contamination create a low enough resistance, the cable may fail immediately. Such a failure can be dramatic and extremely dangerous at any voltage.

**Internal.** All insulation materials can absorb moisture to some degree. As the moisture content increases, the insulation resistance and dielectric strength tend to decrease. Excessive moisture levels can cause the insulation to fail completely.

Moisture content is a special problem with liquid insulation used in transformers and circuit breakers; however, it can cause failure in solid insulation as well.

## Chemical

Some insulation is adversely affected by chemicals. Acids, salts, alkalis, oils, and other hydrocarbons can cause some cable insulation to crack or distort. Chemical reaction can also lower the insulation dielectric strength.

## Dirt

Dirt, dust, and other impurities become a problem to insulation when they are mixed with moisture. The resulting mix of dirt and moisture creates conductive paths. This is a special problem if the contamination is located near the end of the cable close to the termination.

# Chapter 2

# Cable Materials and Construction

## KEY POINTS

- How are low- and medium-voltage cables constructed?
- What types of materials are used for electrical cable conductors? What are their characteristics?
- What types of materials are used for electrical cable insulation? What are their characteristics?
- What types of materials are used for cable jackets? What are their characteristics?

## INTRODUCTION

Cable construction and materials are determined by a variety of factors including:

- Application voltage
- Load current
- Ambient temperature
- Ambient moisture
- Environmental chemicals
- Installation configuration

The history of cable manufacturing and application is a history of changing requirements and innovation. In the early days, the industry began with simple paper and cloth insulation, but now the industry uses synthetic materials. Along the way, paper, oil-impregnated paper, varnished cambric, and other such insulations were (and in some cases are) extremely popular.

Copper has long been, and still remains, the "king" of the conductors. Relative low cost, comparative chemical stability, and low resistance all combine to make copper the ongoing favorite. When metallurgically combined with other materials such as silver, copper becomes even better suited to the job.

Of course, the materials used are only part of the picture. The materials must be formed into a usable, rugged, reliable package that will carry the electric energy to the desired location and keep the electricity out of undesired locations.

## TERMS

The following terms and phrases are used in this chapter:

**Ampacity**: The amount of current that a conductor will carry without exceeding an allowable temperature rise. The ampacity is specified at a given set of ambient conditions.

**Annealed**: To heat and slowly cool to toughen and reduce brittleness.

**Circular mil:** The area of a circle that is 1 mil (0.001 inch) in diameter. Cable sizes are usually given in kcmils; see definition of kcmil (MCM).

**Coated Wire**: Copper wire that has been coated with a tinning compound. Coated wire was originally developed to make copper more impervious to acids, which were used to prepare the wire for rubber coating.

**Ductile**: Capable of being drawn out, as into wire, or being hammered thin.

**Extrude:** To make a material and give it certain properties by forcing it through a die in a liquid or semi-liquid state.

**Hard-drawn**: Conductors that are drawn through dies after they have cooled from the heat process.

**Hygroscopic**: Readily absorbing or retaining moisture.

**kcmil (MCM):** One thousand circular mils; the current preferred designation is kcmil and not MCM.

**Low Voltage:**  a. IEEE definition: Less than 1000 Volts AC.

b. *NEC*® definition: Less than 2000 Volts AC. This definition applies to cable and cable construction requirements. This handbook will use the *NEC*® definition unless otherwise stated.

**Malleable**: Capable of being shaped or formed, as by hammering or pressure.

**Medium Voltage:**  a. IEEE definition: 1001 Volts to 100,000 Volts AC.

b. *NEC*® definition: 2001 Volts to 35,000 Volts AC. This definition applies to cable and cable construction requirements. This handbook will use the *NEC®* definition unless otherwise stated.

**No-oxide**: A chemical treatment put on a metal to prevent oxidation.

**Ozone**: A form of oxygen with three atoms per molecule ($O_3$) rather than two ($O_2$). Ozone is much more chemically active than $O_2$, and is destructive to many types of insulation. Ozone is formed by high energy electrical discharges such as arcing or corona.

**Soft-drawn:** Conductors that are annealed by treating with heat and cooled to remove internal stress.

**Thermoplastic:** An insulating material that is soft when heated and hard when cooled. Examples include polyvinyl chloride (PVC) and polyethylene (PE).

**Thermosetting:** An insulating material similar to thermoplastic with better thermal characteristics. Thermosetting material does not soften as easily as thermoplastic and is not as prone to permanent damage when heated. Cross-linked polyethylene (XLP) is a thermosetting insulation.

**Uncoated Wire:** Wire that has not been coated with a tinning compound.

**Vulcanization:** A process that increases the strength, elasticity, and resistance of a material. It usually involves using sulfur or other additives along with heat and pressure.

## CONDUCTORS

### Conductor Types

Conductors are grouped into two basic types: solid and stranded. A solid conductor is made of a single strand of hard-drawn or soft-drawn wire. Hard-drawn conductors are wires that are mechanically drawn through dies after they have cooled from the heat process. Hard-drawn wire has a high tensile strength and least amount of elongation under stress. This type of conductor is used primarily in outdoor applications such as transmission lines. Hard-drawn conductors are very stiff, and hard to bend and work.

Soft-drawn conductors are wires that are annealed by treating with heat, and cooled to remove internal stress by making the metal less brittle. The soft-drawing process actually softens the metal (copper or aluminum) until it can be easily worked. Soft-drawn wire has low tensile strength and the greatest elongation under stress. Soft-drawn conductor wires bend easier than hard-drawn types and are used for indoor applications.

Solid conductors are primarily used in lighting, service, appliances, and some types of control systems. They are also used in some grounding systems. Solid conductors are difficult to work with in sizes larger than No. 8 AWG. Solid conductors in No. 12 AWG and below work well on screw terminals.

A stranded conductor is composed of single, solid wires twisted together. These single strands of wire may be twisted in a common bundle or individual groups.

Stranded conductors are used more frequently than solid conductors because they are very flexible. It is easier to use them to make connections, such as terminating at device connections and terminal boards. Extra flexibility can be obtained by using many strands of fine wire. Stranded conductors are easy to work with in all sizes, and they ensure good connection.

Stranded conductors also offer slightly less resistance than an equivalently sized solid conductor.

## Conductor Materials

**Introduction.** Table 2-1 lists the materials and characteristics of the more commonly used conductors. The primary materials in use today are copper and alumi-

**Table 2–1  Conductor materials.**

GOOD CONDUCTORS

Silver
- best conductor
- used for contacts
- plated onto some conductors

Copper
- good conductor
- most widely used

Aluminum
- not as good as copper
- strong, light, and cheap
- used for transmission conductors
- used for power distribution

FAIR CONDUCTORS

Brass, zinc, iron, and nickel
Tungsten (used for filaments)
Nichrome (used for heating elements)
Mercury (Used for thermostats and switches)

num. These two materials are used because they are reasonable in cost and are good electrical conductors.

**Copper.** Copper is the most important and commonly used of all conductor materials. Copper is used because it possesses several desirable characteristics.

- Copper is highly conductive (both thermally and electrically). It is second only to silver and gold in conductivity.
- Copper is plentiful and relatively inexpensive when compared to silver and gold.
- Copper is both ductile and malleable. These properties make it ideal for use as a conductor.
- Since it is resistant to both corrosion and fatigue, copper may be used in a variety of industrial and commercial environments.

**Aluminum.** Aluminum is the second most popular material used in the fabrication of electrical conductors. It is cheaper and lighter than copper and has almost as good thermal and electrical conductivity. Unlike copper, aluminum does not possess high tensile strength. In transmission lines, aluminum is reinforced with steel to give it the tensile strength required. Aluminum possesses several other characteristics.

- Aluminum expands and contracts on copper terminals causing high resistance and resultant heat. High resistance builds up because the copper and aluminum have different thermal coefficients; therefore, the terminals loosen when aluminum is used with copper.
- Aluminum corrodes when connected to copper conductors because of galvanic chemical action caused by the reaction of the two dissimilar metals. No-oxide chemical compounds must be used to prevent this reaction.
- Aluminum is used in power systems, depending upon the design and code requirements.
- Aluminum is used in transmission, switchyard, building, and service cable.
- Aluminum does not conduct as well as copper and, therefore, must have a slightly larger cross sectional area for the same ampacity. In spite of this, aluminum conductors are generally

lighter and less expensive than copper conductors of the same ampacity.

**Combination Materials.** In wiring applications where high strength is required, a number of other conductors are available for use. These include coated copper, cadmium copper, chrome copper, and zirconium copper, and are discussed as follows.

- When copper wire is coated with a tinning compound, it is easier to solder and is more impervious to corrosion. Coated wire was originally developed to protect copper from the acid compound that was used to prepare it for covering with rubber insulation.
- Nickel combined with copper strengthens the conductor, makes soldering easier, and reduces corrosive effects. Copper-coated nickel conductors may be used in power, control, and lighting systems.
- Copper covered steel is also available as either a fully annealed or hard-drawn material, the latter being preferred where high conductor strength is required. The conductance of these wires is generally specified as either 30 or 40 percent of an equivalently sized copper conductor at low or medium frequencies.
- Zirconium coated copper provides excellent high temperature performance.

**Silver and Gold.** Where maximum conductance (minimum resistance) is required, silver is widely used. Silver is the best conductor of *all* naturally occurring materials. Since it is quite expensive, it is used in limited applications such as silver-plated contacts.

Although silver is an excellent conductor, it is quite soft and subject to corrosion and oxidation. Because of this, gold is often used in applications where silver is not acceptable. Since it is so expensive, gold is used in smaller electronic applications. The following summarizes the characteristics of these metals when used in electrical conducting service.

- Silver and gold are relatively expensive.

• Both materials are used primarily for contacts and other such limited applications.

• Gold is used when the corrosive tendencies of silver will cause problems.

## Conductor Sizing

Conductors are sized with a wire gauge from high numbers down to size 0 (read "aught"). Size 0 is referred to as 1/0, pronounced "one-aught." Large numbers indicate small sizes of wire. For instance, a size 36 conductor is very, very fine, about the size of a strand of hair. A size 12 conductor is as large as the lead of a standard pencil.

Most conductors smaller than 1/0 can be measured with an American wire gauge (AWG), as illustrated in Figure 2-1. Conductors larger than 1/0 are sized with multiple zeros that increase with size; therefore, a 3/0 conductor is larger than a 2/0 conductor. Conductors are measured with a micrometer caliper. The wire size is usually identified on the insulation or jacket covering.

Conductors larger than 4/0 are measured and sized directly in mils. A mil is a unit of wire measurement equal to one-thousandth of an inch (0.001 inch). A circular

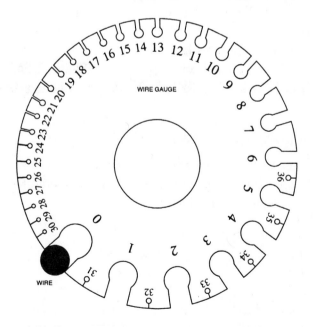

**Figure 2-1   American Wire Gauge (AWG).**

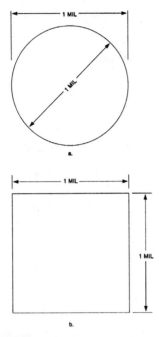

**Figure 2–2 Conductor sizing: a. circular mil (CM) and b. square mil (SM).**

mil is an area measurement equal to the area of a circle with a 0.001 inch (1 mil) diameter. A square mil is the area of a square that is 1 mil on a side. Figure 2–2 shows these two measurements.

Practical values for wire sizes fall into the thousands of circular mils category; for example, 250,000 circular mils or 250 kcmil. (Note: The preferred term for one thousand circular mils is kcmil. The older MCM is no longer used.)

## Current Carrying Capacity (Ampacity)

Ampacity refers to the amount of current a conductor can carry without exceeding an allowable operating temperature. Excessive temperature rise can result in several negative consequences including:

- Conductor melting (fusing)
- Insulation degradation and failure
- Fire

Ampacity is influenced by many factors such as conductor material and size, ambient temperature, installation method, number of conductors in close proximity, and actual current levels.

Ampacity tables are available to determine what size cable may be used for a given application. Refer to the current edition of the *NEC®* for cable ampacity tables.

# INSULATION

## Classifications

Insulation can be classified by its application in various environments. For example:

- Chemical resistant insulation is classified to corrosive and acid environments.
- Heat resistant insulation is classified in terms of use in extreme or abnormal ambient temperatures.
- Moisture resistant insulation is classified for use in areas where moisture is constantly present.
- Oil resistant cable insulation is classified for use where oil is present in unusual or damaging conditions.

Table 310–13 in the 1990 *NEC®* lists all of the types of insulation and identifies the trade names, maximum operating temperature, application provisions, sizes of conductors, thickness of insulation in mils, and outer covering, if any.

## Rubber

Rubber is one of the most common types of insulation. To prevent chemical reaction, it is usually separated from the conductor by various coatings and materials. Rubber is resistant to moisture and is not well suited for use with high temperatures (above 205 degrees Celsius) or high voltages.

Various *NEC®* grades such as Type R, RH, and RHW are applied to rubber insulating materials to withstand high moisture or high temperature conditions and other abnormal application conditions. Rubber is commonly used for appliance cords, portable cords, conductor leads, and portable cables.

## Elastomers

Elastomers are materials that can be compressed, stretched, or deformed and still snap back to their original shape. In this respect they are like rubber. Four elastomers in fairly common use in modern insulation include: Neoprene, Hypalon, Ethylene Propylene Rubber, and Styrene Butadiene Rubber.

**Neoprene.** Neoprene is one of the most important elastomeric materials in use. It has been used successfully as a cable jacket for about fifty years. Where service conditions are unusually abrasive, neoprene is highly recommended. In extreme environments, the neoprene construction is often reinforced with additional layers of cotton or other reinforcing members midway through the jacket wall. Not inherently flame retardant, neoprene jacketed cables must be compounded with the necessary flame retarding chemicals. Like the other members of the elastomeric family, neoprene jackets are usually more expensive both to produce and purchase.

**Hypalon (Chlorosulfonated Polyethylene).** Hypalon possesses the majority of the properties of neoprene, plus it exhibits better thermal stability and resistance to both ozone and oxidation. Like neoprene, it is a fairly expensive material and requires continuous vulcanization (CV) for its fabrication.

**Ethylene Propylene Rubber (EPR).** Ethylene propylene rubber has been used in power cables for quite some time and has just recently been employed in fabrication of telecommunication and other types of wire and cable. When used as a jacketing material, EPR exhibits excellent weathering properties and is quite resistant to ozone. Unlike neoprene, EPR can be compounded to exhibit excellent electrical characteristics. Although EPR possesses good chemical and mechanical properties, it is not inherently flame retardant. Therefore, as in the case of some other jacketing materials, it must be compounded with flame-retarding additives.

**Styrene Butadiene Rubber (SBR).** This is a general purpose synthetic rubber characterized by relatively low tensile strength when stretched. In order for this material to be suitable for use as insulation, the base elastomeric material must be judiciously compounded with other materials, such as carbon black. The SBR materials are known for their excellent abrasion resistance and reasonably good electrical characteristics. However, when exposed to ozone, oil, or weather they readily decompose. These compounds have a useful operating temperature range of about −55 degrees Celsius to +90 degrees Celsius. The material is highly flammable unless steps are taken to compound the base materials with adequate flame retardants.

## Thermoplastics

Thermoplastic is another commonly used type of insulation that is much stronger than rubber. They are often used for conductor insulation and protective covering in cables rated below 2000 Volts. Thermoplastic materials can be softened by heat and flow under mechanical pressure, and maintain their altered shape when cooled. Thermoplastic materials are more rigid than elastomers, have better electrical properties, and are lower in cost. They are also lightweight. The types of thermoplastic used primarily in insulation are described in the following paragraphs.

**Polyvinyl Chloride (PVC).** PVC is the most popular of the thermoplastics. PVC is a strong, water-resistant substance that may be compounded with many materials. Some compounds improve its mechanical attributes, while others accentuate its electrical characteristics. PVC possesses a high resistance to most acids, alkalis, and oils, and can be processed with flame inhibitors to pass most flame tests.

PVC materials have been compounded to perform well at temperatures as low as –55 degrees Celsius and as high as +105 degrees Celsius. Formulations are available that allow PVC materials to yield good service life when used outdoors.

**Polyethylene (PE).** Polyethylene is used whenever low line loss is required. It has excellent electrical properties and is highly resistant to water, ozone, and oxidation. It has, however, low resistance to flames, gasoline, and solvents.

## Thermosetting

Thermosetting insulation is similar in appearance to thermoplastic, but is somewhat thicker. This insulation is cross-linked polyethylene (XLP). It exhibits good resistance to heat, chemicals, moisture, and ozone. It does not soften in normal operating conditions, but will distort at temperatures above 90 degrees Celsius. This insulation is used in high voltage applications, such as motor leads, where temperatures do not exceed 90 degrees Celsius, and in high-voltage cables.

## Nylon

Nylon is a member of the polyamide class of resins and was originally developed for use as filament strands and synthetic fibers.

While nylon is not generally employed as a primary insulation, it is used as a secondary insulation and jacketing material where a high degree of abrasion resistance and or heat and chemical resistance is desirable. Typical thermal resistance of an insulated conductor sheathed with nylon is 105 degrees Celsius.

Nylon is not used as much as other insulation types because it has a strong affinity for moisture, and has relatively poor electrical characteristics.

## Teflon®

Teflon is used as an insulation where high temperatures, varied chemicals, and moisture conditions exist. It also resists the destructive effects of oils. Teflon is used in temperature applications of up to about 200 degrees Celsius. It is used with wiring on appliances, boiler controls, fire alarms, transformer leads, and motor leads.

## Mineral Insulation

Mineral insulated cable is made of highly compressed magnesium oxide and covered with an outer metallic sheath, usually copper or alloy steel. Mineral insulation is designed to be as noncombustible as possible. It is used in feeders, services, and branch circuits in wet or dry locations.

## Paper

Paper by itself is a mediocre insulation. However, when impregnated or immersed in other materials, paper insulation provides the best electrical characteristics of all currently used insulations.

**Solid.** Solid paper insulation consists of layers of paper tape impregnated with mineral insulating oil. A tightly fitting lead sheath covers the insulation system and provides the best mechanical and chemical protection. Lead-covered solid paper insulation is being phased out in favor of newer, less expensive, and more easily used insulation systems.

**Gas-Filled.** This cable is constructed in a manner similar to solid paper insulation. Before the cable is leaded, the oil is drained into a nitrogen atmosphere. Gas feed channels are built into the cable to give the nitrogen free access to the conductors.

**Oil-Filled.** This cable is similar to solid insulation, except that the oil used is a relatively thin liquid, which is fluid at all operating temperatures.

## Varnished Cambric

Varnished cambric is a smooth, yellowish-brown insulation that is being replaced by more modern types. It is composed of cotton cloth coated with insulating varnish, which is fabricated into tape. An oily compound is applied for waterproofing and to prevent friction between the insulation layers when the cable is bent.

Varnished cambric insulation has high dielectric strength, but breaks down when the temperature is above 85 degrees Celsius. It was used in high voltage, high temperature applications such as generator leads, transformer leads, and substations. It is not frequently used in newer installations.

# JACKETS AND SHIELDS

Many cables are covered with a protective jacket for additional thermal, chemical, mechanical, and environmental protection. If the cable has a metal shield (see construction details later in this chapter), the jacket is usually applied directly over the shield. Otherwise, it is placed directly on the insulation.

Most of the insulation materials already discussed may be used as cable jackets. The most popular jacketing materials employed include PVC, PE, neoprene, hypalon, and EPR. Polyurethane and thermoplastic elastomer (TPE) jackets are also gaining popularity.

## Polyvinyl Chloride (PVC)

This is the most popular jacketing material. PVC resins may be compounded in many ways to afford normal mechanical protection in cables suited for indoor applications. Other vinyl compounds find widespread application in the area of process control cables. Process control cables require extremely rugged construction to withstand harsh operating conditions. PVC jackets are compounded with other chemicals for indoor or outdoor use in a wide range of temperatures: from –60 degrees Celsius to +105 degrees Celsius. The rugged PVC jackets used in instrumentation cables are also compounded with flame retardant chemicals so that they have an extra measure of flame retardancy. Because of the additional compounding requirement, this type of PVC jacket is generally more expensive than those used in the more general purpose PVC jacketed wires and cables.

PVC jackets exhibit relatively good resistance to abrasion and the absorption of water. However, in places where constant standing water is expected, or where there are regulatory agency requirements with which the cable must comply, PVC jacketing is not preferred.

## Polyethylene (PE)

PE cable jackets are preferred over PVC jackets where constant water conditions are expected as PE is much more impervious. PE materials are not naturally flame retardant, however, they can be made so by compounding with additives. Even when compounded, they are not as resistant to fire as PVC. Typical service life for polyethylene jacketed cable is over twenty years.

## Polyurethane

Polyurethanes possess excellent abrasion resistance and excellent low temperature flexibility. This family of compounds may be extruded by conventional means without vulcanizing. Polyurethanes are hygroscopic and flammable.

## Thermoplastic Elastomers (TPE)

The TPE family of materials has found increasing application as both a primary insulating material and a jacketing material. Thermoplastic elastomer materials possess excellent environmental and chemical characteristics, which make them a natural for jacketing applications.

## Metal-clad Cable

Metal-clad cable consists of one or more separately insulated conductors twisted together with cords of jute that run the full length of the cable. The conductor assembly is then wrapped with a binding tape or enclosed in a thermoplastic jacket. Next, the entire assembly is enclosed in a metallic sheath. The sheath may be smooth and seamless, welded and corrugated, or interlocking. The interlocking sheath is in the form of interlocking metal tape armor similar to Type AC armored cable.

There are three types of cable in this category: MC, ALS, and CS. These types differ in their construction and in the material used for the metal jacket that encloses the cable.

Metal-clad cables are designed for service voltages of up to 15,000 Volts. Cables designed for 600-Volt service have one or more stranded grounding conductors running their full length.

**Type MC.** This type is enclosed in a corrugated metallic sheath or in interlocking metal tape made of galvanized steel, aluminum, or bronze. The metallic sheath may have a plastic covering to protect it from corrosive conditions.

**Type ALS.** Type ALS metal-clad cable is enclosed in a smooth, seamless aluminum sheath.

**Type CS.** This jacket is comprised of a smooth, seamless copper or bronze sheath.

## Lead Sheath

Lead sheath is used where cable will be buried directly in the ground. Lead is also used where cables will be subjected to water, corrosion, or abrasion. It is sometimes used with an armored jacket.

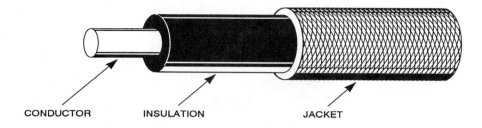

**Figure 2–3  Low-voltage cable construction.**
*(Courtesy of Cadick Professional Services)*

## CONSTRUCTION

### Low Voltage (Less than 2000 Volts)

Figure 2–3 shows the general construction features of low-voltage cable. The center conductor is surrounded by an insulation layer which provides the electrical insulation. The insulation may be covered by a protective jacket. Note that while the protective jacket may have some electrical insulating ability, the actual insulation characteristics of the cable are vested completely in the insulation layer.

The materials used in this system were described previously in this chapter.

### Medium Voltage (2001 Volts – 35,000 Volts)

Because of the greater electrical stress, medium-voltage cable is more complex in construction than low-voltage cable. Medium-voltage cable is composed of five or more individual layers, each of which has a specific purpose. The following explains the six most commonly used layers. Each of the layers is diagrammed in Figure 2–4.

**Conductor:** Made of copper, aluminum, or other conducting material, this layer is responsible for the actual conduction of the electrical current.

**Semiconductor:** This layer is also called the strand shield or strand screen. Air voids between the conductor and the insulation allow the buildup of very high-voltage gradients. The strand screen "short circuits" these voltage gradients and prevents the formation of corona.

**Insulation:** This is the electrical insulation layer. As with low-voltage cable, 100 percent of the electrical insulating capability of the cable is vested in the insulation layer.

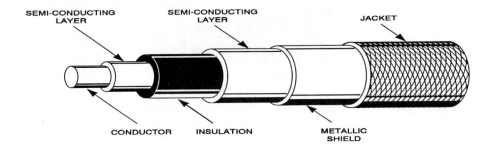

SEMI-CONDUCTING LAYER   SEMI-CONDUCTING LAYER   JACKET

CONDUCTOR   INSULATION   METALLIC SHIELD

**Figure 2-4   Medium-voltage cable construction.**
*(Courtesy of Cadick Professional Services)*

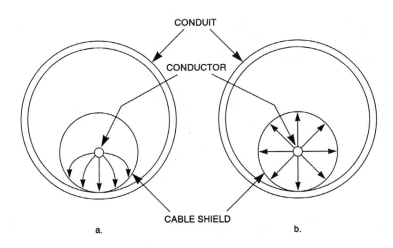

CONDUIT

CONDUCTOR

CABLE SHIELD

a.                                b.

**Figure 2-5   Electric fields inside cables: a. without shield and b. with shield.**
*(Courtesy Cadick Professional Services)*

**Semiconductor:** This layer is similar to the strand shield and serves a similar purpose. This outer semiconducting shield also helps to enclose completely and equalize the electric field within the cable. This will be explained in greater detail shortly.

**Metallic Shield:** This layer serves as an electric shield, which contains and equalizes the electric field inside the cable. Consider Figure 2-5; with no shield the field would be unevenly distributed, as shown in Figure 2-5a. This causes unequal and excessive electrical stress and can lead to insulation failure. Figure 2-5b shows how the shield equalizes the electric field. Under these conditions, the cable life will be greatly extended. Note that this is not usually a problem at low voltages; therefore, cable for low voltages is not typically designed in this manner.

**Jacket:** The jacket provides environmental protection for the cable. Chemical, ultraviolet, and moisture damage is kept from the insulation layer by the jacket. The jacket also protects against mechanical damage.

# Chapter 3

# Cable Installation, Splicing, and Termination

## KEY POINTS

- How can wire and cable be installed?
- What are the advantages and disadvantages of the various installation methods?
- How are wire and cable spliced and terminated?
- What are the differences between medium-voltage and low-voltage splicing and termination?

## INTRODUCTION

Wire and cable are subject to constant attack from electrical, mechanical, chemical, and environmental elements. While the use of proper materials is the key starting point to the long-term survival of wire and cable, proper installation is also a major factor. The best equipment will not survive if it is not installed, spliced, and/or terminated properly. The type of wiring installation depends upon the environment, whether it is to be aboveground or underground, the temperature, and the conductor's size and type. Figure 3–1 shows the various types of installations and their applicability to various situations.

The *NEC*® and other standards have many regulations governing acceptable installation for wire and cable. Considerations such as moisture, temperature, and mechanical stress are considered. This chapter will review many of these items, as well as covering some workmanship concerns.

PERMITTED LOCATIONS

| NEC ® ARTICLE | INSIDE BUILDINGS | OUTSIDE BUILDINGS | UNDERGROUND | CINDER FILL | EMBEDDED IN CONCRETE | WET LOCATIONS | DRY LOCATIONS | CORROSIVE LOCATIONS | SEVERE CORROSIVE LOCATIONS | HAZARDOUS LOCATIONS | MECHANICAL INJURY | SEVERE MECHANICAL INJURY | EXPOSED WORK | CONCEALED WORK | SERVICES OUTSIDE |
|---|---|---|---|---|---|---|---|---|---|---|---|---|---|---|---|
| 346 RIGID STEEL CONDUIT, ZINC-COATED | P | P | P | E | P | P | P | P | P | P | P | P | P | P | P |
| 346 RIGID STEEL CONDUIT, ENAMELLED | P | X | X | X | E | X | P | P | X | P | P | P | P | P | X |
| 348 ELECTRICAL METALLIC TUBING | P | P | P | E | P | P | P | E | P | E | P | X | P | P | P |
| 350 FLEXIBLE STEEL CONDUIT | P | E | E | E | E | E | P | E | X | E | P | X | P | P | E |
| 354 UNDERFLOOR STEEL RACEWAYS | P | - | - | - | P | - | P | X | X | X | - | - | X | P | - |
| 356 CELLULAR STEEL RACEWAYS | P | - | - | - | P | - | P | X | X | X | - | - | - | P | - |
| 362 STEEL WIREWAYS | P | E | X | X | X | P | P | X | X | X | P | X | P | X | X |
| 364 STEEL BUSWAYS | P | E | X | X | X | E | P | X | X | X | P | X | P | X | E |

| 351 LIQUIDTIGHT FLEX, CONDUIT | LIMITED TO USE FOR CONNECTION OF PORTABLE EQUIPMENT, OR FOR MOTORS WHERE FLEXIBILITY IS |
| 374 AUXILIARY STEEL GUTTERS | NEEDED. LIMITED TO USE AS SUPPLEMENTAL WIRING SPACE FOR SERVICES, ETC. |

P - PERMITTED.    E - EXCEPTION.    X- NOT PERMITTED.    - CONDITIONS DO NOT APPLY.

**Figure 3–1   Types of installations.**

## TERMS

The following terms and phrases are used in this chapter.

**Conduit**: A tube or trough for protecting electrical wires and cables. It may be a solid or flexible tube in which insulated electrical wires are run.

**Fish Tape**: A long, flexible tape used to thread and pull cable through conduit.

**Penciling**: Tapering insulation into the shape of a pencil. This is normally done as part of a splicing or terminating procedure.

**Splice:** The physical connection of two or more conductors to provide electrical continuity.

**Termination**: The method used to prepare a cable or wire for connection to other equipment. A proper termination allows for uninhibited current flow and adequate insulation. A termination differs from a splice in that the other equipment to which the wire or cable connects is *not* a continuation of the wire or cable.

## UNDERGROUND CONDUIT SYSTEMS

Underground conduit systems are also known as duct systems and consist of one or more conduits spaced closely together. These systems are often buried directly in the ground in concrete casing. Figure 3–2 shows the general layout for such a system.

The concrete casing is buried in a trench below the earth's surface. The depth of the conduit bank is determined by the voltage and type of service. Duct lines terminate in underground vaults called manholes. Conduits may be made of several materials, including fiber, polyvinyl chloride (PVC), vitrified tile, polyethylene, styrene, and monolithic concrete. The following paragraphs will discuss the two most commonly used materials, fiber and PVC.

### Fiber Conduit

Fiber conduit is made of a wood pulp impregnated with a bituminous pitch. The different sections of the conduit are connected by self-aligning joints. Fiber conduit may be either Type 1 or Type 2. Type 1 must be encased in concrete while Type 2 may be buried directly.

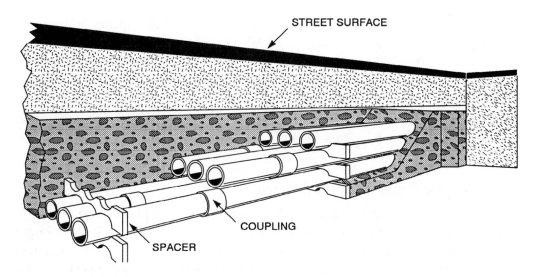

Figure 3–2   Underground conduit system.

Fiber conduit is resistant to corrosion and chemicals such as acids or alkalis. It has a high crushing and tensile strength, and is impervious to moisture. Fiber conduit is not suitable for exposed installations.

## Polyvinyl Chloride (PVC)

Several plastic materials are successfully used for conduit, including PVC, polyethylene, and styrene. Of these, PVC has proven to be the most popular. PVC is low in cost, easy to install, and comes in lengths of up to 30 feet.

PVC is first installed in banks and tiers with appropriate spacers, and concrete is then poured around it. PVC may be used both underground and aboveground, depending upon the requirements of the installation.

# ABOVEGROUND CONDUIT SYSTEMS

Metal or nonmetal conduits may be used in exposed or concealed locations aboveground. Such conduits are made in a variety of materials and configurations.

## Electrical Metal Tubing (EMT)/Thinwall

Electrical metal tubing has a thin steel wall; thus, it is called thinwall conduit. The outer surface is galvanized, while the inner surface is protected with a zinc or enamel coating. The coating on the inside provides a smooth surface for pulling wire or cable.

This conduit is lightweight and ranges from 1/2 inch to 4 inches in diameter. Thinwall conduit cannot be threaded, therefore special box connectors and couplings must be used. The connectors and couplings are secured with set screws, compression denting, or by force fit.

Thinwall conduit is vulnerable to corrosion and rust; however, special types are available with greater resistance to these conditions. Figure 3–3 compares thinwall conduit with rigid conduit. Figure 3–4 illustrates a variety of fittings used with EMT.

## Intermediate Metal Conduit (IMC)

Intermediate Metal Conduit falls between EMT and rigid conduit in terms of weight, strength, and cost. It is made of aluminum or steel and is similar to rigid conduit in many ways. IMC can be threaded and provides a more rugged installation than EMT; however, it has a thinner wall than rigid conduit and should not be used if the installation is subject to severe mechanical stress.

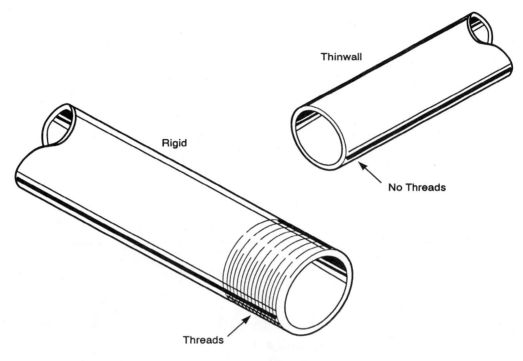

**Figure 3–3   Conduit comparison.**

## Rigid Conduit

Rigid conduit (Figure 3–3) is the most commonly used of the metal conduit classes. Rigid conduit is available in sizes from 1/2 inch to 6 inches in diameter and is made of aluminum or steel. Rigid conduit is usually threaded for installation.

**Galvanized Steel.**  This type is suitable for indoor or outdoor use and can also be embedded in concrete. Galvanized steel is easily threaded, very strong, and has good corrosion resistance. It is also relatively heavy and costly, and corrodes easily if the galvanized coating is penetrated. Since galvanized steel is magnetic, it is a cause of magnetic power loss and poor heat dissipation. Galvanized rigid conduit is also called "white" conduit.

**Enameled Steel.**  Enameled steel (black) conduit is coated with a black enamel. This conduit is commonly used for general interior wiring.

**Sheradized Steel.**  Sheradized steel (green) conduit has a special corrosion resistant coating and is used in installations where high levels of corrosion are expected.

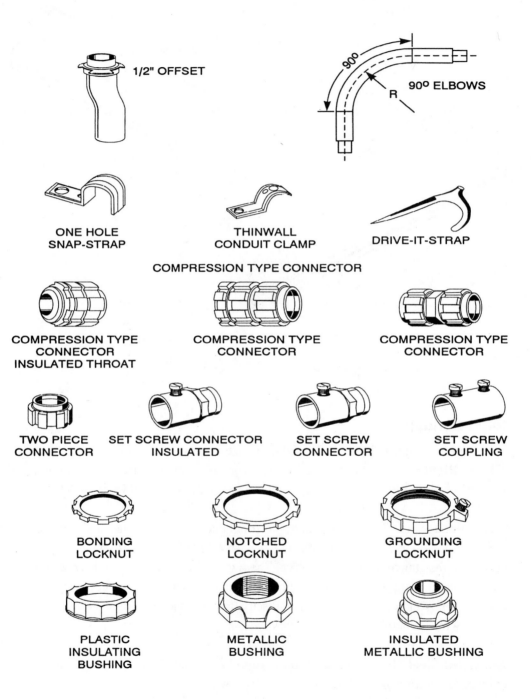

**Figure 3–4   Thinwall conduit fittings.**

**Aluminum.** Aluminum has been increasingly used for exposed indoor and outdoor locations. It is 1/3 the weight of galvanized steel and is noncorrosive in most atmospheres. Aluminum is not as strong as steel, corrodes in some soils, and reacts with concrete.

## Polyvinyl Chloride (PVC)

PVC has a number of advantages and is used in both aboveground and underground systems. The inner walls are smooth and facilitate long cable pulls. The heavy wall type may be found in corrosive or chemical environments. PVC is nonporous and relatively impervious to moisture. PVC is not used in concealed spaces of combustible construction or to support fixtures or other equipment. This conduit is nonmagnetic and nonconducting.

PVC tends to be less expensive than other types of conduit and may be found in type A thin wall, schedule 40 heavy wall, and schedule 80 and 120 extra heavy wall. Schedule 40 has an inside diameter similar to that of rigid metallic conduit. Sunlight resistant types of PVC are also available.

PVC is coupled together using bonding adhesives or glue. Sizes range from 1 inch to 6 inches in diameter. Ten-, twenty-, and thirty-foot lengths are available. PVC has relatively low mechanical strength and is susceptible to heat damage for temperatures above 120 degrees Celsius.

## Liquidtight Flexible Nonmetallic Conduit (LFNC)

LFNC is used where the extra flexibility and corrosion resistance of a nonmetallic conduit is required. LFNC is constructed in three different types.

- A smooth seamless inner core covered with reinforcing layers bonded together.
- A one-piece assembly consisting of a smooth inner core with built-in reinforcement.
- A corrugated inner and outer surface with no reinforcement.

LFNC is made of PVC or some other similar waterproof, corrosion resistant insulating material.

## Flexible Metal Conduit

Flexible metal conduit is made of a single strip of galvanized steel or aluminum wound into a longitudinal spiral and interlocked to form a continuous flexible tube.

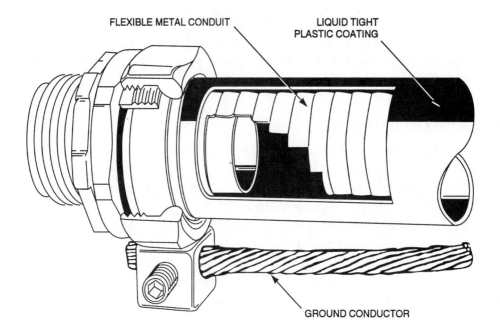

FLEXIBLE METAL CONDUIT                LIQUID TIGHT
                                      PLASTIC COATING

GROUND CONDUCTOR

**Figure 3–5    Liquidtight flexible conduit.**

The resulting construction has many of the advantages of solid metal conduit, in addition to being lightweight and very easy to install. Lengths of up to 250 feet are available. Except for liquidtight conduit (discussed shortly), flexible metal conduit may not be used in wet environments unless the enclosed wire is rated for the location. Bending radii must be closely observed to prevent pinching the cable insulation.

When flexible metal conduit is enclosed in a plastic sheath, liquidtight flexible metal conduit is created. This type of conduit is especially useful for motor lead connections and other installations which require great flexibility and may be subject to water problems. Liquidtight conduit may not be used where physical damage is likely to occur. Figure 3–5 shows a cross section of liquidtight flexible metal conduit.

## Wireways and Wire Ducts

Wireways are sheet-metal troughs with hinged or removable covers for housing and protecting wires and cables. These systems are often used when equipment and control devices are located close to each other.

## Cable Trays

A cable tray is an assembly of units made of metal or other noncombustible materials, which form a continuous, rigid support for cables; see Figure 3-6. Cable trays are used throughout industry, and they greatly simplify the installation of wire and cable. There are four different types of cable trays.

**Ladder Type.** A ladder-type cable tray is a prefabricated metal structure consisting of two longitudinal side rails connected by individual transverse members, which provide the cable supporting means.

**Trough Type.** This is a prefabricated metal structure greater than four inches in width. It consists of a ventilated bottom and has closely spaced cable supports within integral or separate longitudinal side rails.

**Channel Type.** Like all cable trays, the channel type is prefabricated. It consists of a one-piece ventilated or solid bottom channel section not exceeding four inches in width.

**Solid Bottom Type.** This is a metal structure that has no openings in the bottom. The cable support is provided by integral or separate longitudinal side rails.

## AERIAL INSTALLATION

The installation of aerial cable, or overhead as it is sometimes called, offers several advantages, including higher ampacity and low maintenance requirements. Several different types of installations are employed, including insulated triplex and quadraplex cables, noninsulated conductors on insulators, and solid or pipe-type bus. The higher ampacity is a result of the excellent cooling properties of free air.

Aerial cables must be able to withstand the forces of high winds, storms, and ice loading. Self-supporting aerial cables may be strung directly from pole to pole, while other types must be lashed to a high strength steel wire called a messenger wire.

### Pole Installations

Overhead installations are often supported by wooden, epoxy glass, or steel utility pole structures. When installed in such a manner, the conductors are supported on insulators, as shown in Figure 3-7. Such a structure may be used to install insulated or noninsulated wire for overhead systems. Other types of pole construction may be used for dead-ends, angles, and corners.

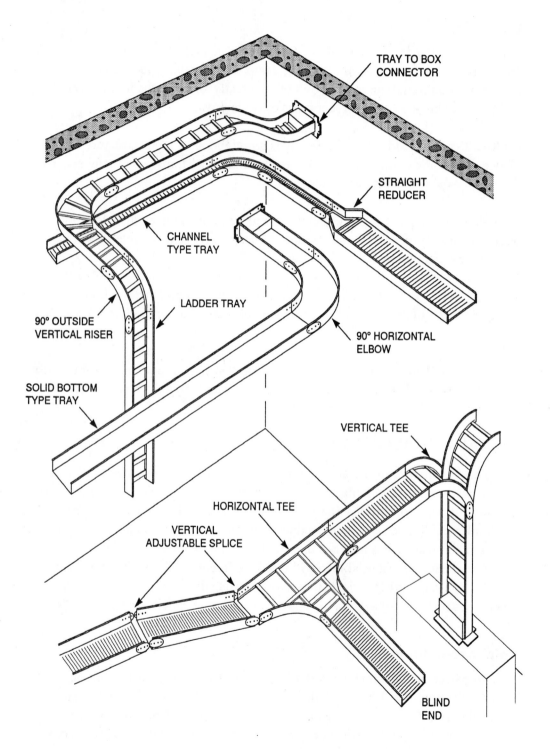

**Figure 3–6   Cable tray system.**

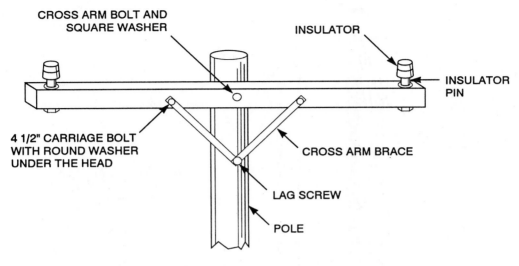

**Figure 3–7   Type of crossarm.**

Some conductor spans are separated by insulating spacers. Such separators hold the conductors apart and stabilize them in high winds.

## Triplex Cable

Triplex cable, as shown in Figure 3–8a, is a self-supporting cable that can be used for overhead runs. It is commonly used for connection of utility distribution circuits to commercial or residential service entrances, as shown in Figure 3–8b.

## DIRECT BURIAL INSTALLATION

Some cables are suitable for direct burial installation. Such cables are placed into the ground by either plowing or trenching. A cable laying plow opens the ground, lays the cable, and backfills the opening in one pass. In the trenching method, a backhoe or trencher is used to dig a trench into which the cable is laid. The trench is then backfilled. Figure 3–9 shows a cable-burying operation.

Buried cables must be protected against frost, water seepage, burrowing animals, and mechanical stresses caused by earth movement. Armored cables specially designed for burial are available. Cables should be buried at least 30 inches deep. This depth puts them deep enough to protect them from damage. In colder states, they are buried below the frost line. Other buried cables should be enclosed in sturdy polyurethane or PVC pipe. These pipes should have an inside diameter several times

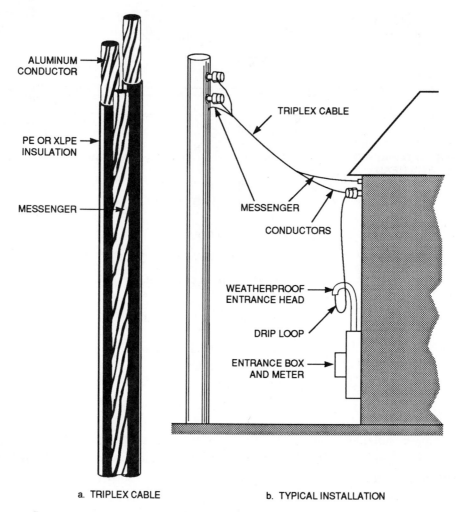

ALUMINUM
CONDUCTOR

PE OR XLPE
INSULATION

MESSENGER

TRIPLEX CABLE

MESSENGER

CONDUCTORS

WEATHERPROOF
ENTRANCE HEAD

DRIP LOOP

ENTRANCE BOX
AND METER

a.  TRIPLEX CABLE

b.  TYPICAL INSTALLATION

**Figure 3–8   Triplex cable and installation.**
*(Courtesy of Cadick Professional Services)*

the outside diameter of the cable to protect against earth movements. An excess
length of cable in the pipe prevents tensile loads being placed on the cable as the
system expands and contracts with temperature variations.

# CABLE PULLING

## Considerations

Cable may need to be pulled into conduit, floor raceway, cable tray, wire mold,
or metal raceway. Certain preparations must be made, regardless of the type cable
used or application. At a minimum, the cable puller should:

**Figure 3–9    Burying underground cable.**
*(Courtesy of Alan Mark Franks)*

- Check the *NEC*® and manufacturer's cable information charts to determine suitability of the cable chosen for the job.
- Check the temperature rating and weight of the cable.
- Survey the environment and terrain.
- Measure the length of the cable run to determine the amount of cable required.

## Insertion of Pull Line

Before a cable can be pulled into a raceway, a pull rope must be in place. Pull ropes are in turn pulled into the raceway by the use of a pull string. The pull string is routed through the raceway using one of three common methods.

**Existing Rope.**  When a cable is pulled into a raceway, a pull rope may be connected to it. The rope is left in place to pull other cables or pull ropes. This method requires care to avoid tangling the rope with the cable as it is pulled in. It is not suitable for small raceways.

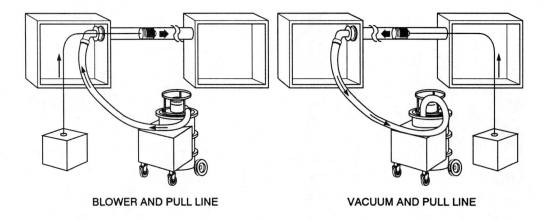

BLOWER AND PULL LINE                    VACUUM AND PULL LINE

Figure 3–10  Using pressure or vacuum to insert pull line.

**Fish Tape.**  Fish tape has long been used to pull rope or string. The tape is "fished" into the raceway and connected to the pull string at the far end. The fish tape is then pulled back with the string connected.

**Pressure/Vacuum.**  Figure 3–10 shows one of the best ways to insert the pull line. The use of a blower or a vacuum requires that the raceway be completely sealed. A lubricated plug is placed into the conduit with the pull string attached.

If a vacuum is being used, the hose is connected to the receiving end of the conduit and the plug/pull wire combination is literally sucked through the pipe. In the case of a blower, the hose is connected to the sending end and the plug/pull wire is blown through the conduit. Note that the blower type requires a special fitting on the hose end to allow passage of the pull wire.

## Rigging

Regardless of whether cable is pulled by hand or machine, it must be properly attached to the pull rope. The method used will be dependent upon the cable size, the length of the pull, the type of wire, manufacturer's recommendations, and other variables. The following sections describe the most common methods used for rigging the cable to the pull rope.

**Taped Connections.**  Short, relatively lightweight pulls may be executed using a fish tape secured to the wire with electrical tape, as shown in Figure 3–11a.

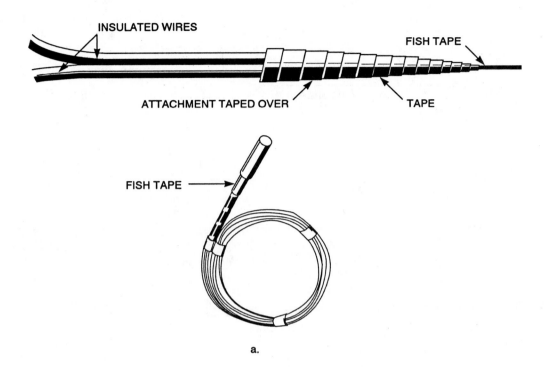

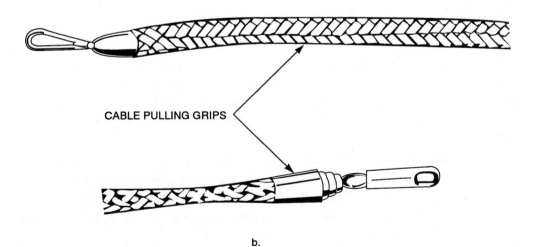

**Figure 3–11    Wire pulling methods:    a. fish tape and electrical
tape connections and b. wire mesh grips.**

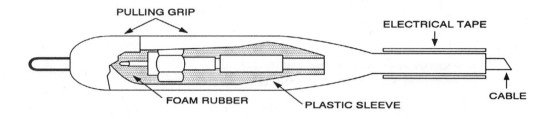

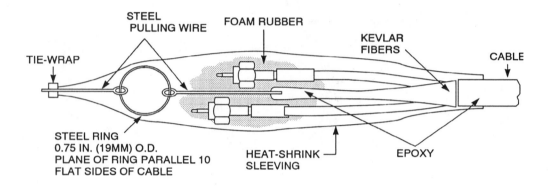

**Figure 3–12   Pulling grips:   a. simplex pulling grip and b. duplex pulling grip.**
*(Courtesy of Canoga–Perkins Corporation)*

**Wire Mesh Grips.** For more difficult pulls, a woven metal pulling grip (Figure 3–11b) may be used. The grip is pushed onto the insulation, with the eye connected to the fish tape or pull rope. The grip mesh is wound in such a way that pulling on it causes it to tighten on the cable. The harder the pull, the tighter the grip.

**Simplex & Duplex Grips.** Several specialty types of grips are available that provide even better pulling characteristics than the common wire mesh types. Figures 3–12a and 3–12b show the simplex and duplex types. These types of grips may grip the sheath as in Figure 3–12a, or may grip support members such as Kevlar fibers contained in the cable itself.

**Compression Pulling Eyes.** Heavy duty pulls can be performed by using a compression pulling eye, as shown in Figure 3–13. This is a special assembly that is compressed onto the cable conductor in much the same way as an electrical termination lug. Some pulling eyes compress onto the insulation as well as the conductor.

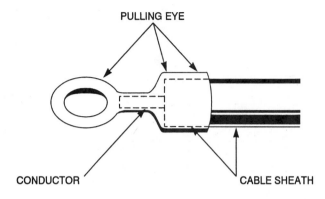

PULLING EYE

CONDUCTOR

CABLE SHEATH

**Figure 3–13   Compression pulling eye.** *(Courtesy of Cadick Professional Services)*

## Accessories

**Swivels.**  As a cable is pulled into a raceway, a tendency to rotate may develop. This tendency usually comes from the pull rope. To prevent this, a swivel (see Figure 3-14) is often used, which is free to rotate. Thus, as the pull rope turns, the swivel rotates and prevents the cable from being twisted.

Some swivels are equipped with a shear pin assembly which will break if the pulling tension exceeds a certain value. Many pulls require the use of this feature to prevent excessive and damaging pull tension.

**Sheaves.**  Sheaves are pulley assemblies that are used to change the direction of the pull force. They are particularly useful for pulling cable into underground ducts that terminate in manholes or other such structures.

Many manholes are equipped with eye bolts, which can be used to mount simple pulley mechanisms. Others are not so equipped, and a guide-sheave rack must be used. Figure 3-15a shows a typical guide-sheave rack, and Figure 3-15b shows an installation of a rack.

When using sheaves, the following factors should be considered.

- The sheaves and the rack must be the proper size for the cable that is being pulled. A sheave that is too small will cause the

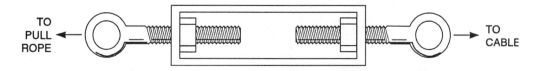

TO PULL ROPE

TO CABLE

**Figure 3–14   Swivel.** *(Courtesy of Cadick Professional Services)*

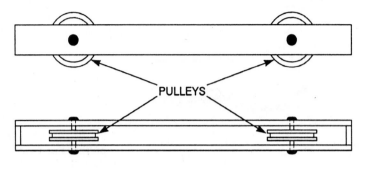

a. GUIDE SHEAVE RACK

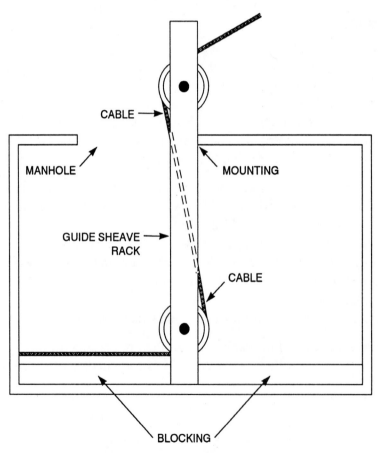

b. SHEAVE RACK INSTALLATION

**Figure 3–15   Typical guide sheave rack and installation.**
*(Courtesy of Cadick Professional Services)*

cable to ride on the pulley ridges and damage the cable. A sheave that is too large will cause the cable to move back and forth and cause an unsteady pull.

- Sheaves cause an increase in the sidewall pressure on the cable during the pull. Refer to the manufacturer's pull tension instructions when using sheaves.

- Be certain to adequately secure the guide-sheave rack. If the rack is inadequately secured, it can slip, causing cable damage or personal injury.

**Protectors.**  Protectors are metal or leather fittings that are placed into the lip of the raceway. As the cable is pulled into the raceway, it rides on the protector and is protected from damage. Manufactured metal protectors may be purchased; yet, many electricians choose to fabricate leather ones in the field.

## Pulling Machines

On longer more difficult runs, cable pulling machines may be employed. Figure 3–16 is a hand-operated pulling machine. This machine operates somewhat like a fishing reel. The cable is connected to the metal cable and then pulled through the conduit by turning the crank on the pulling machine.

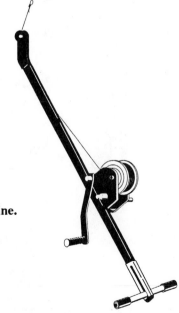

**Figure 3–16   Cable pulling machine.**

For extremely difficult pulls, the motorized cable pulling machine shown in Figure 3-17 may be used. The motorized machine (see Figure 3-18) is started by mounting it onto a stable pipe or stud using the anchoring chains. The rope line is then wrapped around the capstan, and the machine is started.

## Cable Preparation

Always refer to the manufacturer's specifications for the particular type of cable and cable pull applied. The following general practices should always be observed.

- Check prints to determine the proper routing for the cable.
- Ensure that the splice and terminating prints are correct for the cable being pulled.
- Check that all necessary materials are available, such as cable pulling tools and equipment, rope, knife, and proper lubricating compounds.
- Check that the pull will occur at a temperature that is suitable for the cable type.

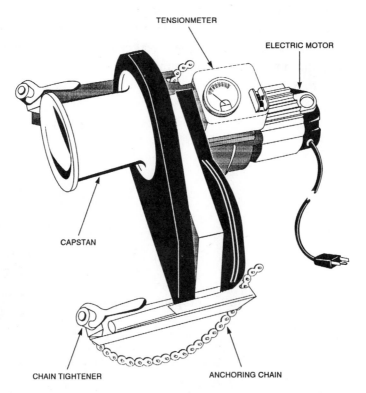

Figure 3-17   Power cable puller.

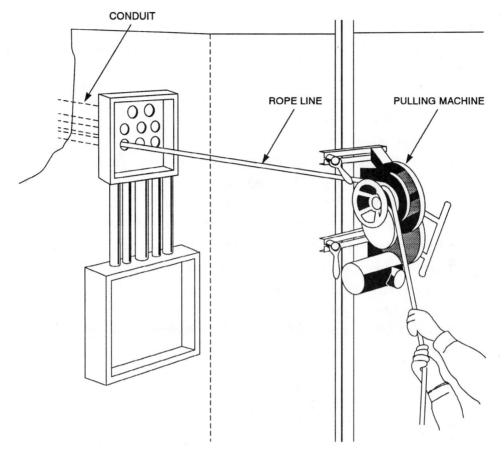

**Figure 3–18  Cable pulling machine in use.**

- Verify that the proper color codes are being used.
- Determine the correct pulling direction.
- For each pull, the cable sidewall pressure must be calculated. Manual calculations or computer programs may be used for this step. Refer to the cable manufacturer or specialty manufacturer for details.
- Position the cable reels properly and watch for any sharp bends or kinks in the cable. Use reel jacks and axles to minimize friction when pulling and feeding the cable.
- Determine the proper number of personnel required for feeding the cable into the conduit.
- Be sure an ample supply of proper lubricant is available.

## Pulling the Cable

The following general practices should always be observed when pulling cable.

- Two-way voice or hand signal communications must be established prior to pulling.
- Short runs may be pulled by hand.
- The cable should be fed as straight as possible into the conduit.
- If basket grips are to be used (see Figure 3–11b), sufficient tension must be applied to get a positive grip on the cable.
- Apply lubricant to the cable constantly during the pull.
- Make certain the sheaves are the proper size.
- Leave some slack if the cable is to be pulled through a pull box.
- Be certain to tag and identify the cable at pull boxes, junction boxes, and terminal points.
- Correct pulling tensions must be calculated *before* pulling starts.

## Cable Pulling Precautions

The following general practices and cautions should be observed.

- Avoid cable crossover where the cables enter conduit and during the pull.
- Observe special care when pulling cables through manholes where damage may occur. Use guide-sheave racks where required.
- Do not allow vertical conduit rises to be supported by conduit fittings (condulets).
- Do not hammer cable into a fitting. Such hammering can severely damage the cable.
- Consider sidewall pressure for vertical hanging cables bent around a corner.
- The direction of the pull and the number of bends affect the tension and sidewall pressure.
- Be certain that the conduit or cable tray is cleared of debris and foreign particles.

- Use a protector inserted in the conduit to avoid abrasion as the cable is pulled in.
- When pulling cable into enclosures with energized conductors, use barriers and guards to be certain that inadvertent contact is not made. Observe OSHA "Safety Related Work Practices" clearance distances.

## Installing Cable in Cable Tray

Cable installed in trays should be pulled over rollers, which are spaced according to the weight of the cable being pulled, as shown in Figures 3–19 and 3–20. These rollers relieve friction and equalize the tension on the cable as it moves through the tray. The following practices should also be observed.

- Use sheaves and cable grips to make the pull.
- Use cover protection during construction.
- Use tension limiting tools if nylon support grips are applied to the cable.

## Tie Offs and Supports

Supports should be provided to prevent mechanical stress on cables where they enter another raceway or enclosure from cable tray systems.

In other than horizontal runs, cables should be fastened or tied off securely to transverse cable trays.

Single conductors should be bound together in circuit groups to prevent excessive movement due to fault current magnetic forces. Some manufacturers require that

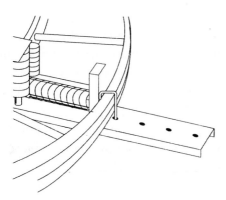

**Figure 3–19  Cable pulling rollers.**

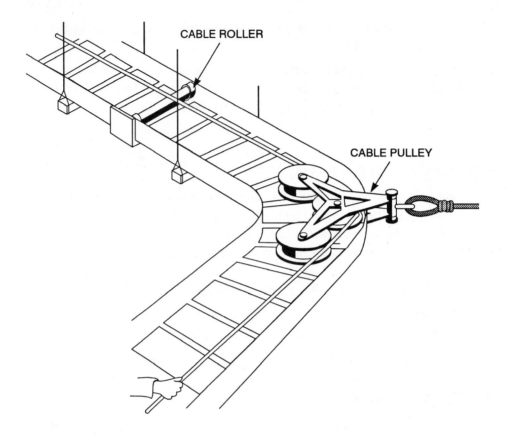

**Figure 3–20   Pulling cable in the cable tray.**

cable be tied off or laced down before equipment will meet Underwriters Labora-tories® (UL®) standards.

## Lubricants

Lubricants must be used to reduce the friction of the cable as it moves through the raceway. Many lubricants are used in the field, including soap, motor oil, and grease. The following sections will discuss four of the most commonly used com-mercially available lubricants.

**Wax/Glycerin/Water Emulsions.** These lubricants provide good friction reduc-tion, but leave messy residues. After the residue dries it may cause the cable to stick in the conduit, making future removal difficult or impossible. Some of the wax lubricants can cause cracking of some cable jackets.

**Bentonite Clay/Water.** These lubricants are slurries of naturally occurring clay in water with or without the addition of ethylene glycol. While they are low in cost and do not damage the cable insulation, they do not provide as much friction reduction as other lubricants.

**Polymer/Water.** Polymer/water lubricants are moderately priced lubricants that are easy to clean. Their unique molecular structure creates an elastic consistency which allows the lubricant to stay with the cable as it pulls through the conduit. They have very low friction and leave little residue.

**Hydrocarbon Grease.** These lubricants are a greasy, jelly-like substance that is obtained from petroleum. While very low in friction, they are quite expensive and extremely difficult to clean from clothing, hands, and cable.

## Pulling Tensions

Calculating maximum and actual pulling tensions must be performed on an individual basis. Always refer to the manufacturer's literature for specific information. In general, the following will apply.

**Maximum Allowed Tension.** The following directions, supplied courtesy of General Cable Company, may be used as a reference for maximum allowed pulling tension.

- The maximum tension should not exceed 0.008 times the circular mil (CM) area when pulled with a pulling eye attached to the copper or aluminum conductors. The formula used is:

$$T = 0.008 \ x \ n \ x \ CM$$

Where:
$T$ = maximum tension in pounds
$n$ = number of conductors in cable
$CM$ = outside diameter of cable in inches

- The maximum stress for lead cables shall not exceed 1500 lb/square-inch of lead sheath area when pulled with a basket grip.
- Maximum tension shall not exceed 1000 lb for nonleaded cables pulled with a basket grip; but the maximum stress calculated

from the preceding formula cannot be exceeded. When pulling a one-conductor (1/C) cable, maximum tension must not exceed 5000 lb. Maximum tension for two or more conductors shall not exceed 6000 lb.

- Maximum tension at a bend must not exceed 300 times the radius of curvature of duct expressed in feet, while not exceeding maximum tension calculated from the first three criteria. Therefore, the minimum radius should not exceed $R(f) = T/300$, where $T$ is the maximum tension as calculated from the first criterion above, or from the manufacturer's information.

**Calculating Tension.** Pulling tension in a horizontal duct may be calculated from the formula:

$$T = L \, x \, w \, x \, f$$

Where:
$T$ = total pulling tension
$w$ = weight of the cable in pounds/foot
$f$ = coefficient of friction

See the manufacturer's literature or cable pulling handbooks for calculating tension in curved ducts and more complicated structures.

## EXPOSED CABLE INSTALLATION

Exposed cable is vulnerable to environmental or physical hazards as it is not protected by metal raceway, conduit, or other enclosures. The *NEC*® requires that conductors be adequately protected where subject to physical damage. The types of protection needed are as follows.

- Protection against penetration by screws or nails by the use of steel plate or bushing of the appropriate length and width.
- When the cable has to pass through wood members, holes must be drilled or notches made.
- Where cables pass through metal framing members and behind panels, the cable should be protected by bushings or grommets securely fastened in the opening.

- Cables routed underground or beneath buildings must be in a raceway which extends beyond the walls of the building.
- Any cables or conductors that emerge from the ground must be protected by an enclosure or raceway.
- Conductors or cables that enter a building must be protected at the point of entrance.
- Cables must also be protected if they are in the immediate area of any activities where people are walking with tools or equipment.

Cables must also be properly secured and supported. Types of supports include clamping devices, tie down straps, cable metal, plastic straps, and cable hangers. Supports must be properly spaced and mounted depending on the size and length of cable run.

Raceways, cable assemblies, boxes, cabinets, and fittings must also be securely fastened in place.

## WIRING INSTALLATIONS IN HAZARDOUS LOCATIONS

A hazardous location is any location where a potential fire or explosion hazard may exist because of the presence of flammable, combustible, or ignitible materials.

The *NEC®* classifies hazardous locations according to the properties and quantities of the hazardous material that may be present. Hazardous locations are divided in three Classes (I, II, and III), two Divisions (1 and 2), four classified Groups (A, B, C, and D), and two unclassified Groups (E and G). Any installations in these areas must be suitable for the locations. Mineral insulated cable (type MI) is the only cable permitted for use in *all* hazardous locations. Other cable types may be used in one or more of the various locations.

Refer to the *NEC®* for detailed information.

## SPLICING

### Overview

Weight and size limitations dictate the amount of cable that may be placed on a reel. Because of this, most installations require that cable be spliced so that runs are long enough. The key feature of a properly made splice is that the splice will have equal or better mechanical and electrical characteristics as the cable being spliced. This section will illustrate several common types of splices.

Although low- and medium-voltage splices generally have the same electrical requirements, the shield on medium-voltage cable must be taken into consideration. Because of this, low- and medium-voltage splices are discussed in separate sections.

## Low-Voltage Splicing and Nonshielded Splicing

**Screw-on Pigtail Connectors.** Screw-on pigtail connectors, also called wire nuts, (see Figure 3–21) are among the most popular and convenient ways to splice wires. Wire nuts are threaded nuts with a plastic insulating cover. To use them, the two or three wires to be spliced are first stripped to a distance of approximately 1/2 inch and then placed parallel to each other. The wire nut is threaded onto the three wires and tightened down in much the same way that a wing nut is used. The resulting connection is both mechanically and electrically sound.

Wire nuts are used in commercial applications for joining fixture wires and branch circuit wires to fixture wires. If the wire nut is not equipped with a metallic coil spring, it may not be used for general-purpose branch circuit wiring.

**Bolted Pressure.** In some lower voltage power applications, bolted pressure connector sleeves are used. Figure 3–22 identifies four basic types of such connectors. Bolted pressure connectors depend upon the applied force of bolts or screws to produce the necessary clamping and contact pressure between the conductors and the connector.

For some connectors, the only tool required for compressing the conductor is a screwdriver. Connectors can be reused and are recommended for use in installations where wiring connections will be changed frequently. Various set screw connectors with thread-on insulated caps are suitable as pressure wire connectors or fixture-splicing connectors, according to individual manufacturer's listings.

The material from which connectors are made varies depending upon the type of wire to be connected. Copper, bronze, and copper/bronze alloys are often used for connectors of copper wire. Aluminum, tin-plated silicon-bronze, or tin-plated copper alloys are used for aluminum wire. Bolted type connections have a few disadvantages, including:

Figure 3–21   Wire connectors.

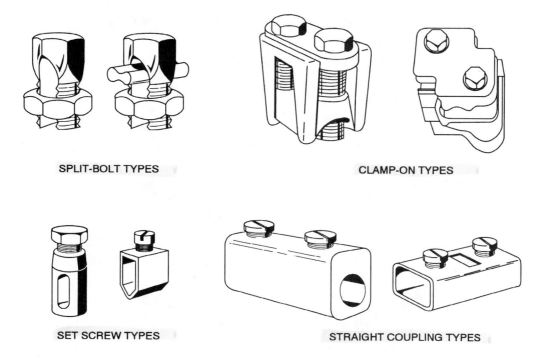

SPLIT-BOLT TYPES

CLAMP-ON TYPES

SET SCREW TYPES

STRAIGHT COUPLING TYPES

**Figure 3–22   Types of mechanical pressure splice connectors.**

- They are subject to the mechanical failure of the screws.
- Bolts may not adequately engage the wire strands.
- Connectors with copper bodies or steel bolts may not give a reliable connection.
- ANSI/IEEE Standard 80 requires that bolted connections be reduced in ampacity. This means that a bolted pressure connector becomes a "weak link" in the chain, and that the cable may not be used to its rated capacity.
- Bolted connectors must be torqued to proper levels before they will meet industry, manufacturer, and code requirements.

**Compression Connectors.** Compression connectors are used throughout the electrical industry. They include all connectors which are applied by a squeezing pressure, either by hand, or by a pneumatic or hydraulic compressing device. Remember that many compression connectors will require the use of a no-oxide lubricant. This is especially important when using aluminum conductors. Refer to the manufacturer's instructions.

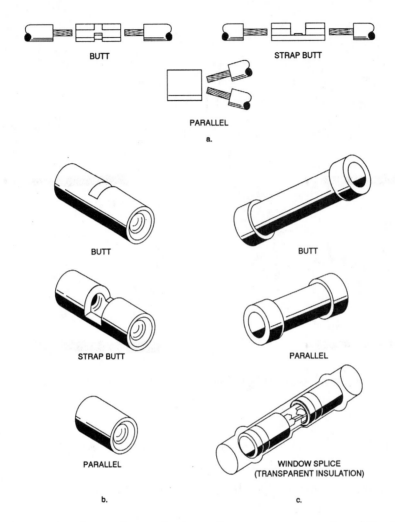

**Figure 3–23**   Low-voltage crimp-on splice connectors: a. crimp splice types,
b. noninsulated crimp-on splice connectors, c. insulated crimp-on splice connectors.

Figure 3-23 shows various types of low-voltage compression-type splice connectors. To use them, a properly stripped length of conductor is inserted into the connector, then a proper compression tool is used to compress the connector. When a wedge-type compression tool is used, the wedge must be placed so that it does not crack the seam on the connector. The connectors shown in Figure 3–23 are all examples of the crimp-on type of compression connector.

Compression connectors as shown in Figure 3–24 can be used for low- and medium-voltage applications. Such a connector is shown properly applied in Figure 3–25.

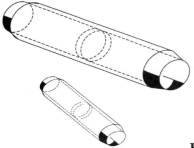

**Figure 3–24    Compression connectors.**

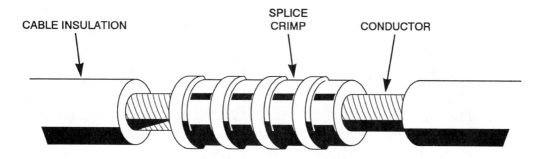

CABLE INSULATION    SPLICE CRIMP    CONDUCTOR

**Figure 3–25    Compressed connector sleeve.**

**Splicing Procedure.**  The most difficult part of any splice is the completion of the insulating system. Figure 3–26 shows the correct procedure for a 0–5kV cable. This procedure applies to polyethylene, cross-linked polyethylene, and EPR insulation. The following procedure refers to Figure 3–26 and is provided courtesy of Rome Cable Corporation.

1. Train conductors into position. Cut ends squarely at centerline of splice.
2. Measure the distance $B + C + D + \frac{1}{2}$ *connector length*, and mark each conductor.
3. Remove jacket and underlying cable tape (if present) from each end of cable.
4. Remove cable insulation for a distance $B + \frac{1}{2}$ *connector length*. Remove conductor shielding tape, if present.
5. Pencil insulation of each end of distance $C$. Smooth with file or garnet cloth, making sure shielding tape (if present) is cut off squarely at ends of pencilled insulation.
6. Round exposed edge of jacket to eliminate sharp corners with a knife, file, or garnet cloth.

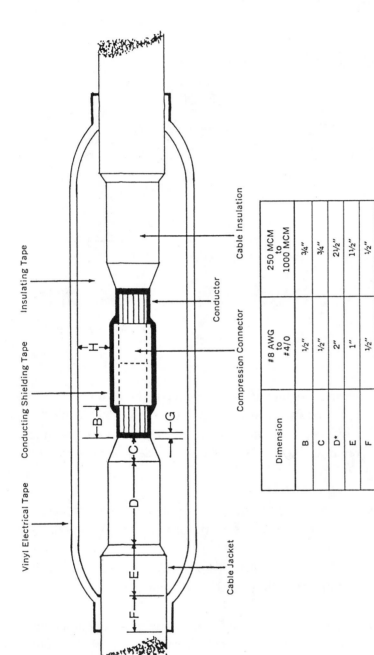

**Figure 3–26    Splicing 0–5 kV cable.** *(Courtesy of Rome Cable Corporation)*

| Dimension | #8 AWG to #4/0 | 250 MCM to 1000 MCM |
|---|---|---|
| B | ½" | ¾" |
| C | ½" | ¾" |
| D* | 2" | 2½" |
| E | 1" | 1½" |
| F | ½" | ½" |
| G | ⅛" | ³⁄₁₆" |
| H | ¼" | ⅜" |

*D = (D + E) when there is no jacket.

7.  Install connector as specified.

8.  Clean jacket, insulation, and connector with a clean cloth which has been moistened with a nontoxic solvent such as trichlorethylene.

9.  Fill connector indents with pieces of conducting shielding tape.

10. Cover connector and exposed conductor area with one half-lapped layer of conducting shielding tape. Note that the tape should overwrap the ends of the pencilled insulation for a distance of 1/16 inch ONLY (dimension *G*).

11. Measure dimension *E* from each end of the jacket and pencil cable.

12. Apply insulating tape to a thickness *H* using uniform tension throughout, and apply in half-lapped layers between the pencils made in Step 11. Stretch the tape to 3/4 of its original width while taping. The diameter of the taped areas over the connector should equal *(2 × H) + (connector diameter) + $^1/_{16}$ inch.*

13. Cover the taped area with two half-lapped layers of vinyl electrical tape. The tape should extend 1/2 inch beyond the insulating tape at each end of the splice.

## Medium-Voltage and Shielded Splicing

The actual joining of the conductors for this type of splice is very similar to low-voltage and nonshielded splicing. The actual procedure comprises five steps: (1) preparation; (2) joining conductors; (3) reinsulation; (4) reshielding; and (5) rejacketing. The following explanation of these steps is provided courtesy of the 3M Corporation.®

**Preparation.**  Begin with good cable ends. A common practice is to cut off the end portion after pulling cable to assure an undamaged end. Use sharp, high quality tools. When the various layers are removed, cuts should extend only partially through the layer. Be careful not to cut completely through and damage conductor strands when removing cable insulation. A useful technique for removing polyethylene cable insulation is to use a string as the cutting tool.

When penciling is required (unnecessary for molded rubber devices), a full smooth taper is necessary to eliminate the possibility of air voids. Semiconductive layer(s) and the resulting residue must be completely removed. Two common methods are to use abrasives and solvents.

A 120–grit abrasive is best for removing semiconductor and residue. A 120–grit abrasive is fine enough for the high-voltage interface, yet coarse enough to remove semiconductor residue without "loading up" the abrasive cloth. Use nonconductive abrasive grit. DO NOT USE emery cloth or any other abrasive that contains conductive particles as these could embed themselves into the cable insulation.

**Figure 3–27 Cable preparation kit.** *(Courtesy of 3M Electrical Products Division)*

Be careful not to abrade the insulation below the minimum specified for the device when using an insulation diameter-dependent device such as molded rubber.

A nonflammable cable cleaning solvent is the preferred method. Avoid using solvents that leave a residue. DO NOT USE excessive amounts of solvent or saturate the semiconductive layers, as this will render them nonconducting. Know the solvent being used, and avoid toxic solvents that are hazardous to health.

There are cable preparation kits available that contain 120–grit nonconductive abrasive cloth saturated with nonflammable, nontoxic 1–1–1 trichloroethane, which make these kits self-contained for field use; see Figure 3–27.

After cable surfaces have been cleaned, the recommended practice is to reverse wrap a layer of vinyl tape, adhesive side out, to maintain the cleanliness of the cable.

High-quality products usually include detailed installation instructions. These instructions should be followed. A suggested technique is to check off steps as they are completed. Good instructions alone do not qualify a person as a cable splicer. Some manufacturers offer hands-on training programs designed to teach proper installation of their products. It is highly recommended that inexperienced splice and termination installers take advantage of such programs.

**Joining Conductors.** After the cables are completely prepared, the rebuilding process begins. The first step is to reconstruct the conductor with a suitable connector. A suitable connector for high-voltage cable splices is a compression-type device; see Figure 3–28. DO NOT USE mechanical-type connectors such as split-bolts. Connector selection is also based on the conductor material.

**Figure 3–28   Compression connectors.** *(Courtesy of 3M Electrical Products Division)*

Generally, copper conductors should be joined with connectors that are marked "CU" or "AL-CU." Aluminum should be joined only with connectors marked "AL-CU." "AL-CU" may come preloaded with no-oxide paste, which is used to break down the aluminum oxide on the surface. If the connector is not preloaded, no-oxide paste must be applied.

It is recommended that a UL-listed connector be used that can be applied with a common crimping tool; see Figure 3–29. This connector should be tested and approved for use at high voltage. In this way, the choice of the high-voltage connector is at the discretion of the user and not limited by the tools available.

**Reinsulating.** Perhaps the most commonly recognized method for reinsulating is the tape method. Tape is not dependent upon cable types and dimensions, and has a history of dependable service and is generally available. Wrapping tape on a high-voltage cable can be time-consuming and error-prone since the careful build-up of

**Figure 3–29   Compression tool.** *(Courtesy of 3M Electrical Products Division)*

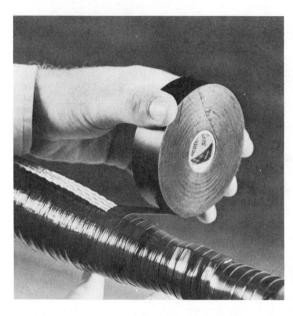

**Figure 3–30 Half-lapping onto a splice.** (*Courtesy of 3M Electrical Products Division*)

tape requires accurate half-lapping and constant tension in order to reduce built-in air voids; Figure 3–30.

New technology has made available linerless splicing tapes. These tapes reduce both application time and error. Studies have shown time savings of 30% to 50% are possible since there is no need to stop during taping to tear off the liner. This also allows the splicer to maintain a constant tape tension, thus reducing the possibility of taped-in voids. Tape splice kits that contain all the necessary tapes, along with proper instructions, are available, and they make an ideal emergency splice kit.

Another method for reinsulating utilizes molded rubber technology. These factory made splices are designed to be human engineered for the convenience of the installer. In many cases these splices are also factory tested and designed to be installed without the use of special installation tools.

All molded rubber splices use EPR as the reinsulation material; see Figure 3–31. EPR must be cured during the molding process using either peroxide or sulphur. Peroxide cures develop a rubber with maximum flexibility for ease of installation, and most importantly, provide an excellent long-term live memory for lasting reliable splices.

**Reshielding.** The cable's strand shield and insulation shield system must be rebuilt when constructing a splice. The tape and molded rubber methods are used in a similar fashion as was outlined for the reinsulation process.

For a tape splice, the cable strand shielding is replaced by a semiconductive

**Figure 3–31    Molded rubber splice kit.**
*(Courtesy of 3M Electrical Products Division)*

tape. This tape is wrapped over the connector area to smooth the crimp indents and connector edges.

The insulation shielding system is replaced by a combination of tapes. Semiconductor material is replaced with the same semiconducting tape used to replace the strand shield, as shown in Figures 3–32 and 3–33.

The cable's metallic shield is generally replaced with a flexible woven mesh of tin-plated copper braid. This braid is for the electrostatic shielding only, and is not designed to carry shield currents. For conducting shield currents, a jumper braid is installed to connect the cable's metallic shields. This jumper must have an ampacity rating equal to that of the cable's shields; see Figures 3–34 and 3–35.

For a rubber molded splice, conductive rubber is used to replace the cable's strand shielding and the semiconductive portion of the insulation shield system; see Figure 3–36. Again, the metallic shield portion must be jumpered with a metallic component of equal ampacity.

A desirable design parameter of a molded rubber splice is that it be installable without special installation tools. To accomplish this, very short electrical interfaces are required. These interfaces are attained through proper design shapes of the conductive rubber electrodes.

Laboratory field plotting techniques show that the optimum design can be obtained using a combination of logarithmic and radial shapes.

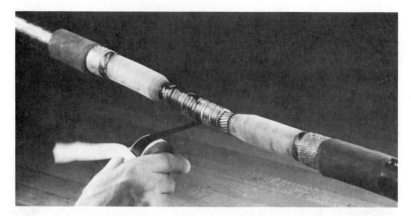

**Figure 3–32   Replacing the cable strand shield with semi-conducting tape.**
*(Courtesy of 3M Electrical Products Division)*

**Figure 3–33   The insulation semiconducting shield.**
*(Courtesy of 3M Electrical Products Division)*

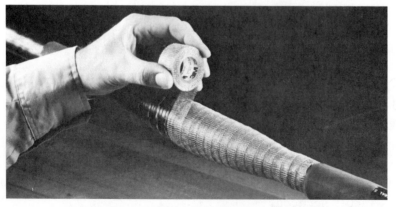

**Figure 3–34   Replacing the electrostatic shield.**
*(Courtesy of 3M Electrical Products Division)*

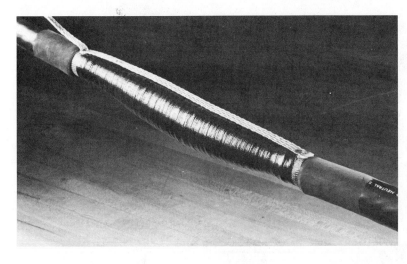

**Figure 3–35
Jumper braid
installed to carry
shield current.**
*(Courtesy of 3M
Electrical Products
Division)*

**Rejacketing.** Rejacketing is executed in a tape splice using a rubber splicing tape overwrapped with a vinyl tape; see Figure 3–37.

In a molded rubber splice, rejacketing is accomplished by proper design of the outer semiconductive rubber, effectively resulting in a semiconductive jacket, Figure 3–38. When a molded rubber splice is used on internally shielded cable (such as ribbon shield or drain wire shield), a shield adapter is used to seal the opening that results between the splice and cable jacket.

**Step-by-step Procedure.** The following procedure is provided by the Rome Cable Corporation, and refers to Figure 3–39.

1. Train cable ends into position. Cut cables squarely at centerline of splice.
2. Measure distance $A$ on each end and mark cables.
3. Remove cable jacket and underlying separator tape from each cable to point marked.

**Figure 3–36
Cutaway view of a
molded rubber
splice.** *(Courtesy of
3M Electrical
Products Division)*

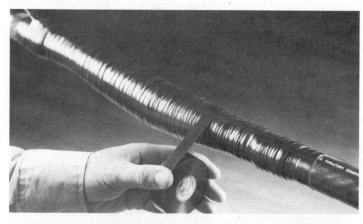

**Figure 3-37　Rejacketing using electrical tape.** (*Courtesy of 3M Electrical Products Division*)

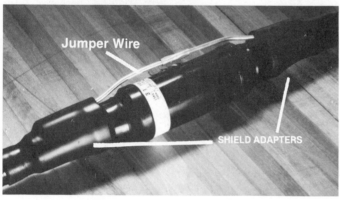

**Figure 3-38　Rejacketing using a molded rubber splice.** (*Courtesy of 3M Electrical Products Division*)

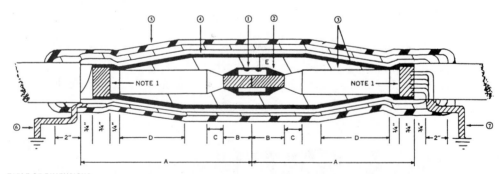

**TABLE OF DIMENSIONS**

| Dimen. | Rated kV | | |
|---|---|---|---|
| | 5 | 15 | 25 |
| A | One Half Connnector Length Plus | | |
| | 6" | 9¾" | 13" |
| B | One Half Connnector Length Plus | | |
| | ½" | ½" | ½" |
| C | ¾" | 1¼" | 2" |
| D | 3½" | 6" | 7½" |
| E | ³⁄₁₆" | ¼" | ⁵⁄₁₆" |

① Compression Connector
② Conducting Tape
③ Insulating Tape
④ Tinned Copper Mesh
⑤ Vinyl Tape, for Plastic Jacketed Cable
　Neoprene Tape, for Neoprene Jacketed Cable
⑥ Ground Lead
⑦ Twisted Drain Wire Ground Lead

**NOTES**

1. Remove all traces of semiconducting material to this point.

**Figure 3-39　Splicing 5, 15, and 25 kV 1/C shielded polyethylene, cross-linked polyethylene, and EPR insulated cables.** (*Courtesy of Rome Cable Corporation*)

4. Remove metal shielding:

   a. For tape shielded cable, remove to where 3/4 inch of this material protrudes beyond the jacket.

   b. For wire shielded cable, unwind wires, bend over jacket, and twist them into a common conductor.

5. Remove the semiconducting layer to the point where 3/4 inch of this material protrudes beyond the metal shielding.

6. Temporarily tie the shielding in place with tape.

7. Remove cable insulation for the distance $B$. Do not nick or cut conductor.

8. Remove strand shielding which should be cut off cleanly at the junction with the conductor.

9. Pencil insulation of each cable end for the distance $C$. Smooth with file or abrasive cloth.

10. Inspect the area at the end of the pencil for protruding fibers of strand shield.

11. Pencil and smooth exposed edges of cable jacket.

12. Install connector per manufacturer's instructions. Smooth any major roughness with a file or abrasive cloth.

13. Wipe jacket, insulation, and connector with a clean cloth, preferably one which has been moistened with a suitable solvent, such as trichloroethane or trichlorethylene.

14. Fill connector indents with pieces of conducting shielding tape.

15. Cover connector and exposed conductor area with one 1/2 lapped layer of conducting shielding tape.

16. Apply insulating tape to a wall thickness of $E$ over the connector shield. Employ level-wind technique and apply in successive half-lapped layers. The ends of the applied insulation should be tapered for a distance $D$ as shown, and should end approximately 1/4 inch from the edge of the cable insulation shield.

17. Remove the tape applied in step 6 holding the semiconducting tapes.

18. Apply a half-lapped layer of conducting shielding tape over the insulating tape. Starting at the very edge of the metal shielding, cover the cable semiconducting layer, the insulating tape, and the semiconducting layer on the opposite side of the splice.

19. Remove the tape applied in step 6 holding the cable metal shielding.

20. Apply copper mesh shielding braid. Start at one of the cable metal shields and apply a half-lapped layer of the braid across the splice to the other metal shield overlapping this metal shielding. Solder the tape to the metal-

lic shielding tape at each end of the braid wrap, avoiding the use of excessive heat.

21. With tape shielded cable, solder a ground lead to the braid, avoiding the use of excessive heat. Solder block this lead or use a solid conductor.

22. Apply jacket tapes. Scrape jacket clean for 3 inches on each side of splice.

    a. For Neoprene jacketed cables, apply three, half-lapped layers of insulating tape across the splice. Extend the insulating tape wrap at least 2 inches beyond the braid at each end of the splice area.

    b. For plastic jacketed cables, apply three, half-lapped layers of insulating tape across the splice. Extend the insulating tape wrap at least 2 inches beyond the braid at each end of the splice area. Next, apply two, half-lapped layers of vinyl electrical tape over the insulating tape, and extending 1 inch onto the cable jacket.

23. Insert shield ground between layers of jacketing tape.

## Other Splice Insulation Methods

A variety of splice insulation methods are available for plastic insulation in addition to the tape wrapping and the molded kit types, which have been discussed. The following paragraphs briefly describe these additional methods.

**Poured Epoxy/Resin-Filled.** This type of insulation (see Figure 3–40) is particularly useful when a splice on a nonshielded cable must be moisture proof. The following steps are executed after the conductors are properly spliced using one of the previously described methods.

1. Snap the two piece mold body together over the splice area.
2. Seal the ends of the mold body using electrical tape.
3. Insert the funnel(s) into the mold body.
4. Prepare the resin and pour slowly into the funnel until filled.
5. After the resin sets, remove the funnels and the mold body.
6. Dress the resin splice as required.

**Heat Shrink.** Heat shrink tubing is another excellent choice for replacement of the insulation and jacket on a low-voltage cable splice; see Figure 3–41. The application of heat shrink is quick, simple, and provides a dependable covering. The procedure is straightforward.

**Figure 3–40    Resin splice insulating kit.**
*(Courtesy of 3M Electrical Products Division)*

1. Slip the heat shrink tubing over one of the ends to be spliced.
2. Splice the conductor as described under low-voltage and nonshielded splicing.
3. Position the heat shrink over the splice and apply heat from a portable torch to shrink the tubing until it conforms to the splice.

The principle disadvantage to heat shrink material is the requirement to use heat. Use extreme care when working with open flame.

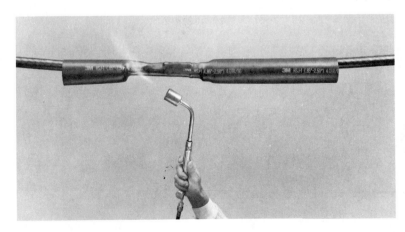

**Figure 3–41    Heat shrink being applied to a splice.**
*(Courtesy of 3M Electrical Products Division)*

**Cold Shrink.** A recent addition to insulating splices is the use of cold shrink materials. Cold shrink tubing is a prestretched, tubular rubber sleeve with a removable, collapsible core. When the tubing is applied in the field, the core material is removed, thus allowing the tubing to collapse, or shrink onto the splice. The procedure is as follows:

1. Slip the cold shrink tubing over one of the ends to be spliced.
2. Splice the conductor as described under low-voltage and nonshielded splicing.
3. Apply a layer of electrical tape.
4. Position the cold shrink tubing over the splice and start removing the collapsible core; the rubber tubing will collapse onto the splice.

Cold shrink tubing creates a heat, chemical, and moisture resistance covering. Figures 3–42, 3–43, and 3–44 show the procedure for applying cold shrink tubing.

**Lead Sheathed Cable.** Medium-voltage lead cable splicing kits are available, which make the task of splicing lead much simpler. Their principle disadvantage is cost. The general procedure for splicing lead, using such a kit, is as follows.

1. Take the preformed lead tube and slide it over one end of the two cables.
2. Complete the conductor splice as previously described. The conductor may be soldered when splicing lead cable, but this is not required.
3. Layer tape onto the splice.
4. Slide the preformed tube over the splice and seal both ends with solder.
5. Fill the tube with the resin compound supplied with the kit.
6. Plug or solder the fill holes.
7. Allow the splice to completely cool before moving it.

Transition kits are also available for splicing plastic cable to lead sheathed cable.

## TERMINATIONS

### Overview

The termination of a cable has much in common with splicing. A cable termination generally needs to provide the following.

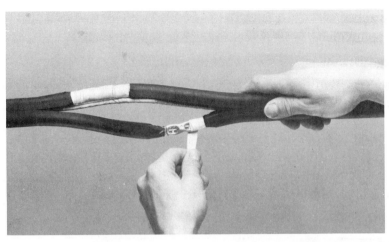

**Figure 3–42    Completing the conductor splice prior to installing the cold shrink.**
*(Courtesy of 3M Electrical Products Division)*

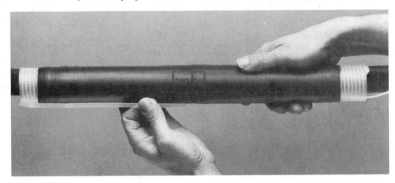

**Figure 3–43    Positioning the cold shrink in starting to remove the collapsible core.**
*(Courtesy of 3M Electrical Products Division)*

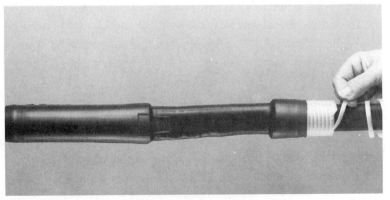

**Figure 3–44    Collapsible core almost removed from one end.**
*(Courtesy of 3M Electrical Products Division)*

- Low electrical resistance connection from the cable to the equipment to which the cable is being terminated.
- Adequate insulation from all conductors.
- Protection from the entrance of moisture, chemicals, and other harmful environmental effects.

The level of sophistication required for a given termination will be dependent upon the voltage level, the current requirements, the size of the cable, and the type of insulation. The following paragraphs will describe several different types of terminations, and will provide some information to execute them.

## Low Voltage

**Terminal Binding.** Figure 3-45 shows a typical terminal binding type of termination. This type of connection is typically used in branch circuit-to-fixture types of connections and control cable terminations. The wire sizes vary, but are usually #10 AWG and smaller. Solid or stranded wire may be terminated in this way, but solid wire is preferable. If stranded wire is used, one of the following methods may be used.

**Low-Voltage Terminal Lugs.** When stranded wire is used, low-voltage crimp-on terminal lugs are the preferred way for terminating low-voltage wires in sizes of #10 AWG and smaller. Figures 3-46 and 3-47 show typical lugs and their application to the wire. Lugs of this type are squeezed on using a hand crimper. Figure 3-48 shows a typical hand crimper.

Note that these types of terminal lugs may be used to allow the use of stranded wire where solid wire is normally employed. For example, solid wire (NM or NMC) is normally used in electrical junction boxes. If stranded wire is used because of its flexibility or higher conductivity, crimp-on terminal lugs must be applied to the wire first. The lug can then be inserted under the screw head or nut and firmly secured.

**Compression Lugs.** Compression lugs are used for the higher current, voltage, and wire size applications. The compression lug benefits from application pressure all the way around the wire, as opposed to the smaller crimped type of connection. Figures 3-49 and 3-50 show compression type lugs and tools respectively. Figure 3-51 shows examples of various types of compression lugs after they have been applied to the wire. This type of lug is used typically on #8 AWG and larger wire sizes.

Lug compressors are designed to fit a certain type of lug; however, some

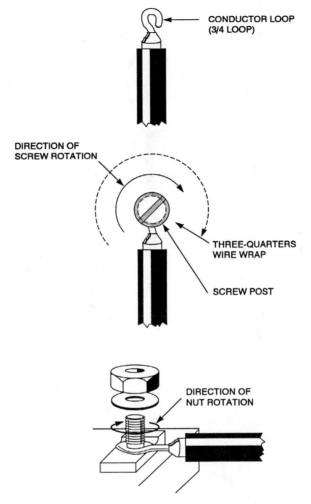

CONDUCTOR LOOP
(3/4 LOOP)

DIRECTION OF
SCREW ROTATION

THREE-QUARTERS
WIRE WRAP

SCREW POST

DIRECTION OF
NUT ROTATION

**Figure 3–45  Typical terminal binding methods.**

compressors are equipped with interchangeable dies to accommodate various types of lugs. Manufacturer's instructions and reference tables *must* be followed to ensure that the correct die is used for the lug being applied.

**Bolted Terminal Lugs.**  Bolted terminal lugs are used for virtually all voltage and current ratings. The principal advantage of this type of termination method is the ease of application and removal. Consequently, bolted terminal lugs are commonly applied in applications where frequent wiring changes may be required.

Bolted terminal lugs are subject to loosening due to heat, vibration, or chemical action. As a result, the lug must be applied carefully with the proper amount of torque

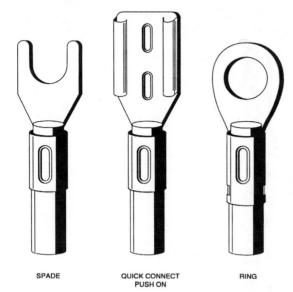

SPADE

QUICK CONNECT
PUSH ON

RING

**Figure 3–46   Typical low-voltage crimp-on terminal lugs.**

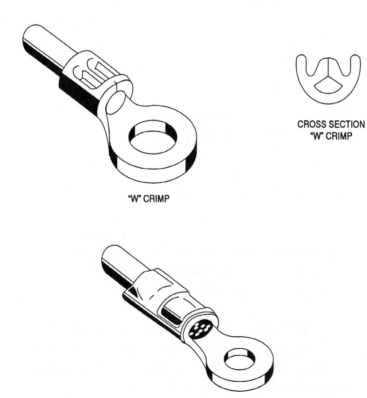

"W" CRIMP

CROSS SECTION
"W" CRIMP

"C" CRIMP

**Figure 3–47   Low-voltage terminal lugs after crimping.**

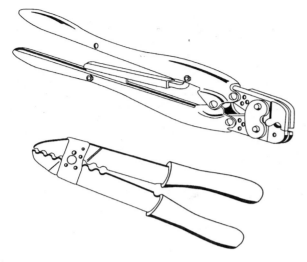

**Figure 3-48    Hand-operated crimping tools.**

**Figure 3-49    Compression lugs.**

administered by a torque wrench. Figure 3-52 shows various types of bolted terminal lugs, and Figure 3-53 shows the application of a bolted terminal lug using a torque wrench.

**Procedure.**  One of the most important steps in any cable termination is the restoration of the insulation system. The following procedure, furnished by Rome Cable Corporation, gives a typical example for a complete termination; see Figure 3-54.

1.  Train cable end into position. Mark and cut off squarely.
2.  Measure a distance 3/4 inch + depth of bore in terminal lug from cable end. Remove cable jacket, insulation, and strand shielded tape (if present) to this point.

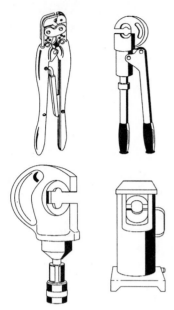

**Figure 3–50   Mechanical and hydraulic compression tools.**

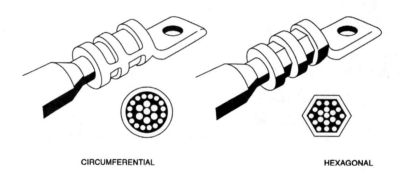

CIRCUMFERENTIAL                                    HEXAGONAL

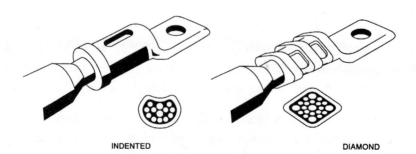

INDENTED                                    DIAMOND

**Figure 3–51   Compression configuration for high-voltage terminal lugs.**

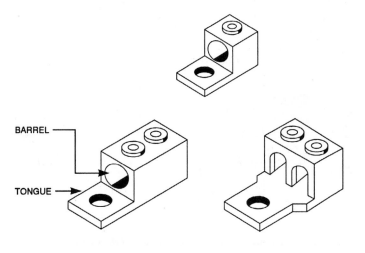

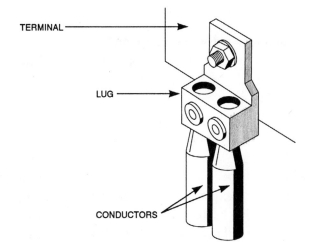

BARREL

TONGUE

TERMINAL

LUG

CONDUCTORS

**Figure 3–52    Bolted terminal lugs.**

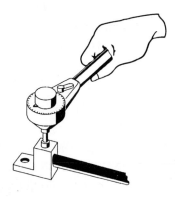

**Figure 3–53    Torque wrench for application of a bolted terminal lug.**

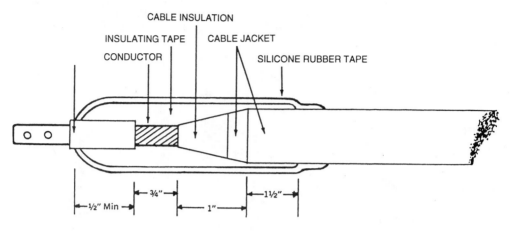

CABLE INSULATION

INSULATING TAPE   CABLE JACKET

CONDUCTOR

SILICONE RUBBER TAPE

¾"

½" Min

1"

1½"

**Figure 3–54   Termination for unshielded 0–5 kV cable polyethylene, cross-linked polyethylene, and EPR insulated cables.**
*(Courtesy of Rome Cable Corporation)*

3. Pencil the cable jacket and insulation to a distance of 1 inch.
4. Smooth the surface of pencilled area with file or cloth. Be sure strand shielding is cut *cleanly* at end of pencil, and that no fibers remain.
5. Install the terminal lug.
6. Wipe the uppermost 2 inches of cable jacket, pencilled area, conductor, and terminal lug with a clean cloth, preferably moistened with a solvent such as trichloroethane or trichlorethylene.
7. Fill indents in terminal lug with filler tape.
8. Apply insulating tape. Start at area of exposed conductor and build up to level of lug shoulder. Apply successive half-lapped layers of tape. Build a tape wall from the broad end of the pencil to a point not less than 1/2 inch up the terminal lug. When this wall reaches a diameter equal to the diameter of the cable jacket, complete the taping by applying an additional three half-lapped layers of tape extending at least 1 1/2 inches onto the jacket.
9. For maximum protection from water and atmospheric contamination outdoors, overwrap the insulating tape with two half-lapped layers of self-fusing silicone rubber tape.

## Medium Voltage

The termination of medium- and high-voltage cable differs from low-voltage cable principally in the method used to alleviate high-voltage electrical stress. Under normal circumstances, the electric field in a shielded cable is uniform along the axis

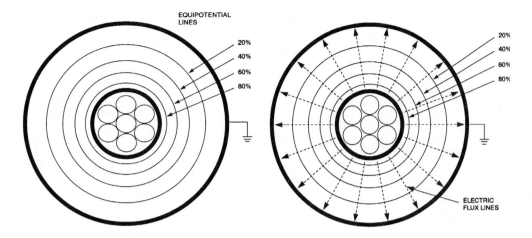

**Figure 3–55   Variation of electric flux in a shielded cable.** *(Courtesy of 3M Electrical Products Division)*

**Figure 3–56   Flux lines in a shielded cable.** *(Courtesy of 3M Electrical Products Division)*

of the cable and varies along the radius of the cable. The greatest voltage gradient is located towards the center (conductor) of the cable. This condition is shown in Figures 3–55 and 3–56.

When a cable is terminated, the semiconductor shield is removed for some distance from the end of the cable. This will cause the electric flux lines to concentrate at the end of the semiconducting shield, as shown in Figure 3–57.

There are two ways to reduce the electrical stress: geometric stress control, and capacitive stress control. In the geometric method, shown in Figure 3–58, the diameter of the termination is expanded and reduces the stress by enlarging the radius of the shield. The method produces the familiar stress cone at the end of many medium-voltage terminations.

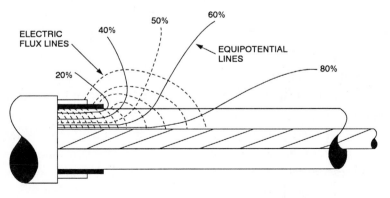

**Figure 3–57   Flux lines in shielded cable cut for termination.** *(Courtesy of 3M Electrical Products Division)*

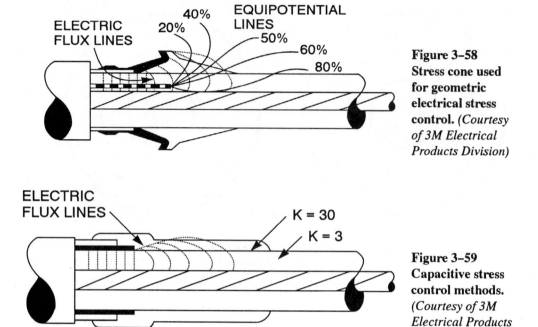

**Figure 3–58**
**Stress cone used for geometric electrical stress control.** *(Courtesy of 3M Electrical Products Division)*

**Figure 3–59**
**Capacitive stress control methods.** *(Courtesy of 3M Electrical Products Division)*

Capacitive stress control methods use materials with different dielectric constants at the termination; see Figure 3–59. In this technique, materials with progressively lower dielectric constants ($K$) are wrapped around the cable. This change in dielectric constant allows the flux lines to distribute evenly, thus reducing the stress to tolerable levels.

High-voltage termination manufacturers provide medium-voltage termination kits that contain all of the necessary materials to terminate, insulate, and seal the termination. Figure 3–60 is an example of such a termination.

## Terminal Blocks

Terminal blocks, as shown in Figure 3–61, are assemblies that are used for splicing and terminating low-voltage wire used primarily in control applications. Figure 3–62 shows the general configuration for connecting an external cable to an internal wiring system.

Terminal blocks may be used to connect either solid or stranded wire. When solid wire is used, the connection may be an eye type of connection, as shown in Figure 3–45, the tubular pressure screw type connector, as shown in Figure 3–63a, or the screw clamp type connector, as shown in Figure 3–63b.

**Figure 3–60    Medium-voltage terminations.**
*(Courtesy of 3M Electrical Products Division)*

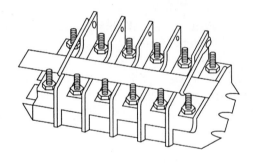

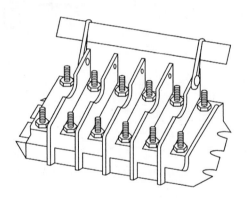

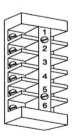

**Figure 3–61    Various types of terminal blocks with marker strips.**

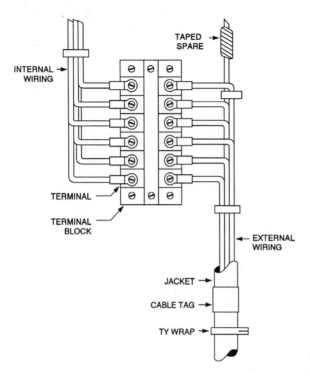

INTERNAL → WIRING

TAPED → SPARE

TERMINAL →

TERMINAL BLOCK →

← EXTERNAL WIRING

JACKET →

CABLE TAG →

TY WRAP →

**Figure 3–62    Wiring connections to terminal block.**

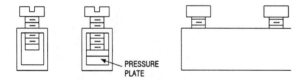

PRESSURE PLATE →

a. TUBULAR PRESSURE SCREW TYPE CONNECTOR

b. SCREW CLAMP TYPE CONNECTOR

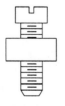

c. SCREW TYPE CONNECTOR

**Figure 3–63    Typical screw-type connectors for terminal blocks.**

When stranded wire is used, the wire is first terminated in a crimp or compression type lug as described earlier. The lug is then placed under the screw type connector (see Figure 3–63c) and the screw is tightened. To insure a secure and lasting connection, split-ring lockwashers or self-locking screw heads should be used on the terminal block.

# Chapter 4

# Cable Testing

## KEY POINTS

- What is the difference between an acceptance test and a maintenance test?
- What is the purpose of electrically testing wire and cable?
- What type of test equipment is used to perform tests on wire and cable?
- What type of tests are performed on wire and cable?
- What specific safety precautions should be employed when testing wire and cable?

## INTRODUCTION

No matter what its age, any piece of equipment is subject to damage. New wire and cable may have manufacturing defects or suffer shipping damage. Service aged wire and cable will be subject to the normal deterioration caused by heat, vibration, chemicals, and electrical stress. Because of this, wire and cable should be tested when new, and thereafter periodically, to determine their serviceability.

This chapter will describe the various facets of cable testing equipment, procedures, and safety precautions. Most cable tests are insulation tests; consequently, the material in this chapter will focus on insulation testing. The reader should note that the material in this chapter is *not* designed to "create" a cable test person. Rather, concepts and theories will be stressed. For detailed training in cable testing, the reader is referred to one of the many training organizations offering hands-on experience.

## TERMS

The following terms and phrases are used in this chapter.

**Absorption Current**: The current that flows as a result of mobile ionic charges in the insulation. As the "capacitor" charges, it attracts ions in the insulation. The motion of these ions creates current flow.

**Acceptance Test**: A test performed on a piece of equipment when it is new or newly installed.

**Apparent Power:** The total voltage multiplied by the total current in an alternating current circuit. The apparent power is measured in units of volt-amperes (va). Apparent power is equal to the vector sum of the true power and the reactive power.

**Capacitive Current:** The current that flows as a result of the charging of the cable "capacitor." For a cable, one "plate" is the center conductor, and the other is the shield or ground system.

**Conduction Current:** The actual leakage current that flows through the insulation. This current is analogous to the current flow through a conductor.

**Continuity Test:** A test that evaluates the condition of an electrical circuit. A circuit that is continuous has a relatively low resistance from one end to the other.

**Ionic Charge:** An electric charge caused by a surplus of positive or negative ions.

**Ions:** A particle, usually subatomic, which has a positive or negative electrical charge.

**Maintenance Test**: A test performed on a piece of equipment that has been in service for some period of time.

**Megohmmeter:** An instrument designed to measure very high resistances, typically over one million ohms. Megohmmeters usually feature relatively high test voltages of 100, 250, 500, 1000, 2500, 5000, 10,000, or 15,000 volts, and relatively high input impedances on the order of 500,000 ohms or more.

**Overpotential Test**: Overpotential Testing, also called high potential testing, implies the use of voltage levels above the nominal or rated voltage of the cable.

**Power Factor:** The ratio of true power to apparent power. When multiplied by 100 power factor, it is referred to as the percent power factor.

**Proof Test**: A test that *proves* the capability of the system to withstand voltage or current. Most commonly used to describe an overpotential test wherein an insulation system is intentionally subjected to more voltage than designed for. If the insulation withstands the proof test, it should be able to withstand the normal system voltage.

**Reactive Power:** The reactive portion of the apparent power. Reactive power is calculated by multiplying the applied voltage times that portion of the current that is 90 degrees out-of-phase with the voltage. Reactive power is measured in units of volt-amperes reactive (var).

**Total Current:** The sum of all the component currents. This is the current that the meter measures. The total current starts at a relatively high value and gradually decreases to a final value equal to the conduction current. A current meter measuring total current flow would start high and gradually reduce to a final value. Inversely, a megohmmeter will start low and gradually increase to a final value.

**True Power:** The resistive portion of the apparent power. True power is calculated by multiplying the in-phase magnitudes of the voltage and the current. True power is measured in units of watts (w).

## PURPOSE

### Acceptance

Problems with new cable or newly installed cable are usually caused by either manufacturing defects or shipment damage. Before cable is placed into service it should be checked to make certain that it is capable of performing correctly.

**Insulation.** Insulation tests are performed to insure that the insulation has not been damaged or manufactured incorrectly. An insulation test evaluates the insulation for excessive leakage currents. Cracks, voids, or impurities in the insulation will cause excessive leakage current, which may result in eventual failure of the insulation. Acceptance tests will isolate these problems before the cable is allowed to go into service, thus preventing an expensive, unplanned outage caused by cable failure.

**Continuity.** When power cable is installed, it must be terminated at each end and may be spliced at intermediate locations. These terminations and splices will exhibit high resistance if they are not properly made. The high resistance will cause excessive heat and ultimately result in insulation failure.

Control cables may include a myriad of wires, each of which must be terminated at the proper location. If any wire is not terminated properly, the control system will fail to operate, causing downtime, troubleshooting expense, or damaged equipment.

Both power and control cable terminations and splices can be evaluated using continuity tests. These tests will spot installation problems before they have an opportunity to cause damage.

### Maintenance

As cable ages, various environmental and service conditions will cause it to deteriorate mechanically and electrically. Vibration can cause insulation layers to slip and crack; heat can cause terminations to loosen; and chemical agents can act on the

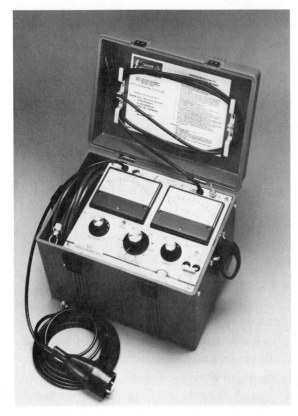

**Figure 4–1   Typical megohmmeter.**
*(Courtesy of AVO Biddle® Instruments)*

insulation and reduce its resistance. Cable and wire, like all other pieces of equipment, have a fixed lifetime. How long they last depends on the environmental and service conditions encountered.

Wire and cable must be evaluated periodically to determine serviceability. Proper evaluation of the test results can provide clues as to the remaining life left in the cable.

**Insulation.**   Vibration, heat, and chemical action all contribute to the aging process of insulation. Periodically, tests should be made to determine the condition of the insulation. By comparing test results at each maintenance interval, the continued serviceability of the insulation can be determined.

**Continuity.**   Heat and vibration are the principal sources of high resistance problems. Thermal cycling will cause expansion and contraction of the termination, which will, in turn, cause the connection to loosen. Vibration aggravates the problem by shaking the connection and accelerating the loosening effects. Worse, as the connection loosens, the resistance and heat increases.

Periodic testing of connections will catch deteriorating connections before they have a chance to fail completely and cause major problems.

## TEST EQUIPMENT

A variety of different types of test sets may be employed for the testing and evaluation of cable systems. The following paragraphs provide a brief description of each one.

### Megohmmeter

The megohmmeter is a direct voltage instrument designed to measure very high resistance values. To obtain sufficient sensitivity, megohmmeters employ relatively high voltages. Commonly used values include 100, 250, 500, 1000, 2500, 5000, 10,000, and 15,000 Volts. Even at these voltages, the high resistance of cable insulation allows only very small currents to flow. Consequently, the instrumentation employed in megohmmeters must be very sensitive.

Figure 4–1 shows a typical modern megohmmeter. Megohmmeters are available with varying degrees of sophistication. The key components of a megohmmeter are the meter(s), voltage control(s), and meter range switch(es).

A megohmmeter will typically have three leads, as shown in Figure 4–2. The line lead is the high-voltage lead that provides direct voltage to the test sample. The

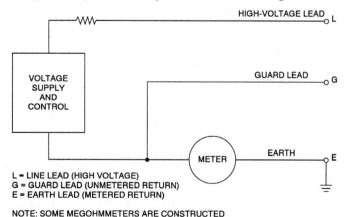

**Figure 4–2   Simplified diagram of a megohmmeter.**
*(Courtesy of Cadick Professional Services)*

**Figure 4–3    Typical hand-cranked megohmmeter.** *(Courtesy of AVO Biddle® Instruments)*

earth lead provides a return path for the current through the insulation specimen. The guard lead is also a return path lead, but it bypasses the meter.

The guard lead is employed when the tester does *not* wish to measure some part of the insulation. For example, if we wish to avoid measuring surface current on an insulator, the surface current will be returned to the instrument through the guard lead. Note that on some instruments, the line lead is metered and the guard lead is at high potential.

Because insulation tests often take place on construction sites where electric service may be difficult to obtain or is poorly regulated, many megohmmeters are equipped with their own built-in, hand-cranked generators. Figure 4–3 is an example of a hand-cranked megohmmeter.

## High-Potential Test Set

High-potential, or overpotential, test sets differ from megohmmeters in several important ways.

- High-potential test sets have higher energy outputs. This translates to greater current capacity when performing a test.
- High-potential test sets are available in much higher voltage ranges than megohmmeters. High-potential test sets with voltage outputs up to 500,000 Volts are available.
- High-potential test set meters are *usually* micro-ammeters in-

stead of megohmmeters. This means that test results are displayed in current values rather than megohms. While the resistance can still be calculated as the ratio of the voltage to the current, this is usually not required.

- Although direct voltage units are the most common, alternating voltage high potential test sets are also available. Alternating voltage units have the advantage that they tend to stress insulation in the same manner that the system voltage does. They have the disadvantage that qualitative measurements are difficult to make. Transformers and certain motor and generator circuits should never be tested with high direct voltage. Cable however, may be tested with high direct voltage *as long as proper test procedures are observed.* See the section on "Types of Insulation Tests" later in this chapter.

- High-potential test sets are normally used to *intentionally* apply more voltage than the insulation rating. In this sense the high potential test becomes a "proof" test. If the insulation is able to withstand the overpotential, it will probably withstand the normal system voltage.

With the exceptions mentioned, high-potential test sets have most of the features of megohmmeters. In fact, high-potential test sets can be used to perform the same types of tests as megohmmeters; yet the megohmmeter is used for certain types of tests because of its smaller size and greater portability. Figure 4–4 shows a typical high-potential test set.

## 0.1 Hertz Test Set

Even though manufacturers routinely apply overpotential to cable insulation for quality control, industry has long questioned the wisdom of intentionally applying excessive voltage to any piece of equipment. Several research studies completed long ago proved that *properly applied* direct voltage overpotential tests will not degrade or decrease the life expectancy of oil impregnated, paper insulated cable.

Some more recent studies, however, indicate that thermoplastic insulation *may* be damaged under the following conditions.

1. A direct voltage overpotential test applies more than twice the rated cable voltage.

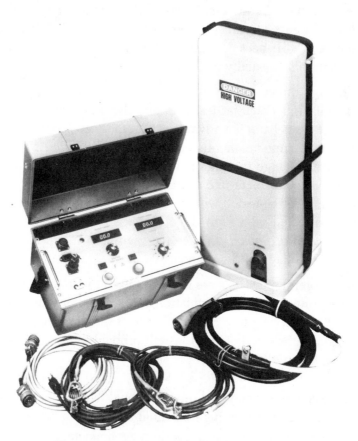

**Figure 4–4   Typical high potential test set.**
*(Courtesy of AVO Biddle® Instruments)*

2. The cable is not grounded for a sufficient time period after the test is completed.

3. The power system voltage is applied at a worst case point on the waveform such that the applied voltage adds to the residual voltage in the cable.

The probability of all three of these events occurring during a test is relatively small, but it is not zero. Consequently, many testers, especially in Europe, have gone to the 0.1 Hertz type of test.

A 0.1 Hertz test set applies a 1/10 Hertz, alternating voltage to the cable. This test has the advantages of DC testing, but can be used in such a way that little or no residual voltage is left on the cable. This test has become increasingly common in Europe, but it has not found widespread support in the United States. Figure 4-5 shows a typical 0.1 Hertz test set.

**Figure 4–5    Very low-frequency test system.** *(Courtesy of AVO Biddle® Instruments)*

## Power Factor Test Set

When an alternating voltage is applied to an insulation system, the resulting current flow is extremely reactive, leading the applied voltage by an angle that approaches 90 degrees. If the insulation deteriorates or is damaged, the resistive current will increase, while the capacitive current remains relatively constant.

Insulation can be evaluated based on the angle between the voltage and current, or power factor, which is the ratio of the real power (watts) consumed by the insulation to the total power (volt-amperes) applied.

A power factor test set, shown in Figure 4-6, is used to determine this angle. Power factor test sets are not often used for evaluation of cable because of the high capacitive current drawn by the cable.

# CONTINUITY TESTS

Continuity tests fall into two basic categories: lead verification and resistance. These two tests, and the equipment required, are described in the following paragraphs.

## Lead Verification

When multi-conductor control cables are being terminated, the electrician must know where each wire end goes. A continuity test is often performed to aid in this operation. The procedure is quite simple.

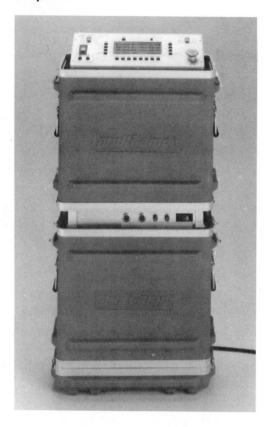

**Figure 4-6   Typical insulation power factor test set.** *(Courtesy of AVO Multi-Amp® Corporation)*

1. One end of the selected wire is grounded or otherwise connected to a known return conductor.
2. A continuity indicator, such as a multimeter or a simple battery-powered buzzer, is connected from the other end of the wire to ground.
3. A continuity indication shows that the selected wire is the correct one. This type of test is commonly performed during the construction phase of an electrical facility. The type of equipment used for this procedure is usually quite simple and often homemade. Figure 4-7 shows a typical buzzer test set used for this purpose.

## Resistance

Often the electrician will need to know the resistance of a connection, which are typically a few tens of micro-ohms. These values can be used to compare against previous readings to indicate a deterioration of the connection. The type of instrument used for such a test is called a micro-ohmmeter or a milli-ohmmeter, depending on the magnitude of the resistance present; see Figure 4-8.

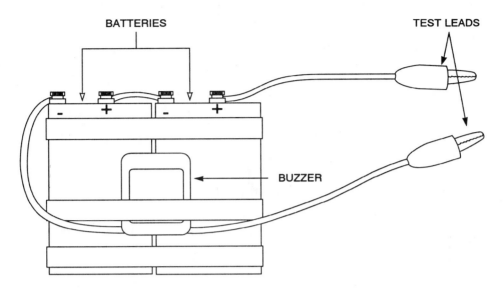

**Figure 4–7  Buzzer test set for continuity testing.**

To use this instrument, the current leads are first placed across the termination or joint that is to be evaluated; see Figure 4–9. The potential leads are then connected *inside* the current leads, as shown. Two lead sets are used so that the measurement is taken from the joint, as opposed to the test set ends of the current leads. In this way the measured resistance includes only the joint, and not the leads or connections.

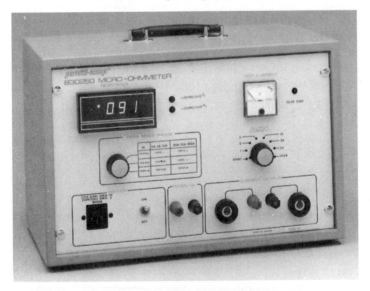

**Figure 4–8  Typical micro-ohmmeter.**
*(Courtesy of AVO Multi-Amp® Corporation)*

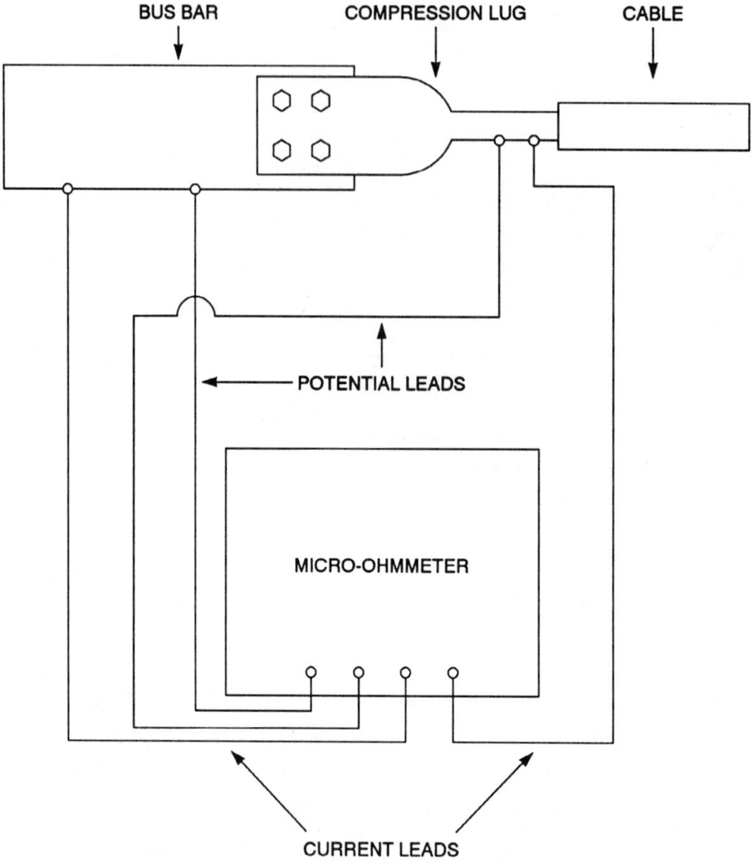

**Figure 4-9   Using the micro-ohmmeter.**
*(Courtesy of Cadick Professional Services)*

Connections can be a significant part of the overall resistance. Current is applied from the test set, and typically ranges from 10 amperes to 400 amperes, with the resistance read directly from the meter.

This type of test is normally done as a maintenance or troubleshooting procedure.

## INSULATION CURRENT

Wire and cable insulation presents a capacitive circuit for the passage of electric current; see Chapter 1 for more information. Figure 4-10 shows a graphical representation of the current that will flow through insulation when a direct voltage is suddenly applied to it. The different types of current flows—conduction, capacitive, absorption, and total—are defined in the Terms section at the beginning of this chapter.

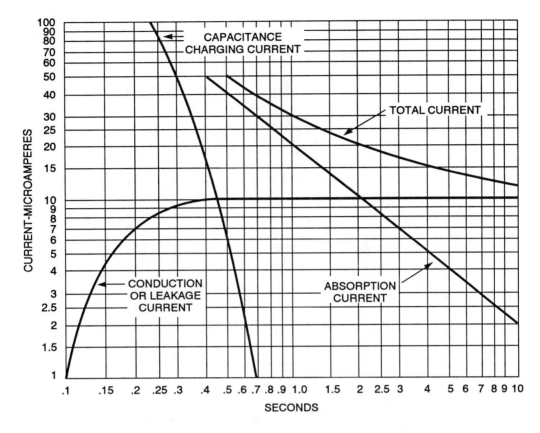

**Figure 4–10** Curves showing components of current measured during DC testing of insulation.

## TYPES OF INSULATION TESTS

### CAUTION:

THE TEST PROCEDURES GIVEN IN THE FOLLOWING SECTIONS ARE FOR REFERENCE ONLY! PERSONS UNFAMILIAR WITH THE OPERATION AND SAFETY HAZARDS ASSOCIATED WITH THIS TYPE OF TESTING SHOULD NOT PERFORM ANY OF THESE TESTS WITHOUT APPROPRIATE SUPERVISION. THESE TEST PROCEDURES ARE NOT INTENDED TO BE USED FOR AN ACTUAL TEST. USE *ALL* APPROPRIATE SAFETY PROCEDURES INCLUDING, BUT NOT LIMITED TO, THOSE THAT ARE LISTED AT THE END OF THIS CHAPTER, MANUFACTURER'S RECOMMENDATIONS, ELECTRICAL CONTRACTOR'S PRACTICES, AND CUSTOMER/OWNER'S SPECIFICATIONS.

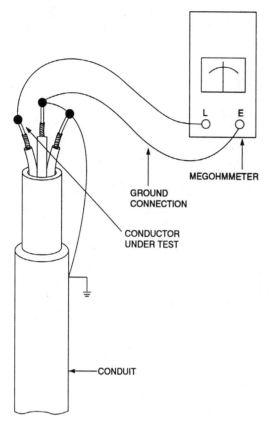

GROUND
CONNECTION

MEGOHMMETER

CONDUCTOR
UNDER TEST

CONDUIT

**Figure 4-11    Cable test.**

## Insulation Resistance

Figure 4-11 shows a megohmmeter connected to measure the insulation resistance of one conductor of a three-conductor cable. This particular configuration will give the resistance of the conductor with respect to ground. The test procedure is as follows.

1. FOLLOW ALL APPROPRIATE SAFETY STEPS FOR THIS TEST.
2. Select a proper voltage range for the test. Low-voltage cable may be tested at either 500 Volts or 1000 Volts. Refer to the manufacturer's recommendations for medium-voltage cables.
3. Ground all conductors that will not be part of the measurement.
4. Connect the earth lead of the megohmmeter to the conductor being tested.
5. Connect the line lead of the megohmmeter to the conductor being tested.
6. Energize the megohmmeter at the voltage selected in step 2.

7. Wait one minute, taking note that the indicated insulation resistance is gradually rising. This rise occurs because the capacitive and absorption currents in the cable are decreasing.

8. Read and record the insulation resistance from the megohmmeter scale after one minute. One minute is chosen as a reasonable time to allow the current to settle.

9. Locate an appropriate temperature correction table for the insulation being tested, and correct the measured resistance for 20 degrees Celsius.

10. Record all measured and corrected values and compare them to previous readings and local standards.

11. If the megohmmeter switch has a drain or ground position on its selector switch, turn it to this position long enough to drain any static charge from the cable; from 1 minute to 1/2 hour depending on the cable size.

12. Reconnect as needed to measure the resistance of the other conductors.


## Polarization Index

A polarization index test is performed using the same setup as for the insulation resistance test; see Figure 4–11. Use the following procedure.

1. FOLLOW ALL APPROPRIATE SAFETY STEPS FOR THIS TEST.

2. Select a proper voltage range for the test. Low-voltage cable may be tested at either 500 Volts or 1000 Volts. Refer to the manufacturer's recommendations for medium-voltage cables.

3. Ground all conductors that will not be part of the measurement.

4. Connect the earth lead of the megohmmeter to the conductor being tested.

5. Connect the line lead of the megohmmeter to the conductor being tested.

6. Energize the megohmmeter at the voltage selected in step 2.

7. Wait one minute, taking note that the indicated insulation resistance is gradually rising. This rise occurs because the capacitive and absorption currents in the cable are decreasing.

8. Read and record the insulation resistance from the megohmmeter scale after one minute.

9. Continue the test an additional nine minutes for a total of ten minutes. Take and record another reading after ten minutes.

10. Divide the ten minute reading by the one minute reading and compare to local standards. Table 4–1 is a table of generally acceptable values for the polarization index.

**Table 4–1   Dielectric absorption ratios.**

| Insulation Condition | 60/30–second ratio | 10/1–minute ratio polarization index |
|---|---|---|
| Dangerous | – | Less than 1 |
| Questionable | 1.0 to 1.25 | 1.0 to 2 |
| Good | 1.4 to 1.6 | 2 to 4 |
| Excellent | Above 1.6 | Above 4 |

11.  If the megohmmeter switch has a drain or ground position on its selector switch, turn it to this position for long enough to drain any static charge from the cable; from 1 minute to 1/2 hour depending on the cable size.

12.  Reconnect as needed to measure the polarization index of the other conductors.

## Overpotential Tests

A direct-voltage overpotential test may be performed on medium-voltage cable for maintenance or acceptance testing purposes. Overpotential testing is not usually performed on low-voltage unshielded cables. Figure 4–12 shows an AVO Biddle Instruments Model 220110 test set connected to perform an overpotential test on a cable. The following explains each of the connections shown in that figure.

The Biddle 220110 DC test set consists of two units. Cable *C1* is the connecting cable between the control unit and the high-voltage transformer unit. Connection *A2* is the line (high-voltage) connection, and *A1* is the guard connection. The guard connection is normally used in these tests as the extremely high-voltage levels will cause surface leakage currents to flow that will interfere with the test results. Cable *C2* contains the 120 volt line cord, the safety foot switch, and the ground connection. Notice that the untested cables are grounded and jumpered together.

The overpotential test is performed by applying voltage to the cable in five to ten equal steps up to a maximum value. At each step, the current is read after the time varying components have settled out; this interval is usually from one to five minutes. When the maximum value is reached, the voltage is left on for a period of fifteen minutes, with current readings taken every minute. The following procedure may be used.

1.  FOLLOW ALL APPROPRIATE SAFETY STEPS FOR THIS TEST.

2.  Select a proper maximum voltage for the test. Refer to Table 4–2 or local engineering standards to determine this voltage. Manufacturer's recommendations will always take precedence over other sources of voltage information. Note that Table 4–2 applies only to new cable. Do not use values for old service aged cable.

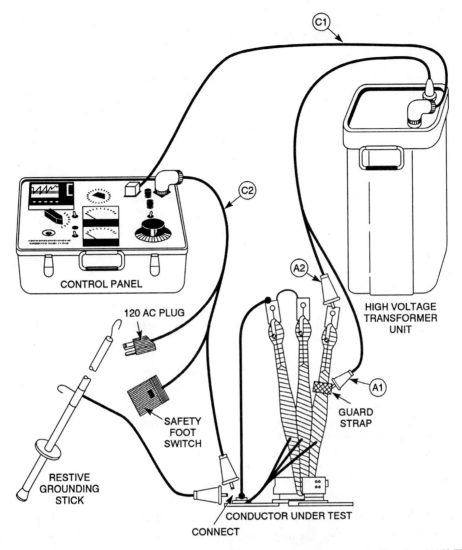

**Figure 4–12   Typical test connection using the 220110 DC Dielectric Test Set (110kV).**
*(Courtesy of AVO Biddle® Instruments)*

3. Ground all conductors which will not be part of the measurement.
4. Connect the earth lead of the test set to the ground.
5. Connect the line lead of the test set to the conductor being tested.
6. Connect the guard lead of the test set to the insulation of the conductor being tested. This connection should be made using a connection strap placed six inches to one foot away from the terminal connections.
7. Determine the number of voltage steps to be used in the test. Divide the maximum test voltage by the number of steps and round to the nearest

**Table 4–2   Maximum voltage for acceptance testing of new cable.** *(Courtesy of NETA Acceptance Testing Specification, 1991, InterNational Electrical Testing Association, PO Box 687, Morrison, CO 80465)*

| Insulation Type | Insulation Level | Rated Cable Voltage | Test Voltage kV dc |
|---|---|---|---|
| Elastomeric Butyl & Oil Base | 100% | 5kV | 25 |
| | 100% | 15kV | 55 |
| | 100% | 25kV | 80 |
| | 133% | 5kV | 25 |
| | 133% | 15kV | 65 |
| Elastomeric: EPR Ethylene-Propylene Rubber | 100% | 5kV | 25 |
| | 100% | 15kV | 55 |
| | 100% | 25kV | 80 |
| | 100% | 35kV | 100 |
| | 133% | 5kV | 25 |
| | 133% | 15kV | 65 |
| | 133% | 25kV | 100 |
| XL, XLP Polyethylene | 100% | 5kV | 25 |
| | 100% | 15kV | 55 |
| | 100% | 25kV | 80 |
| | 100% | 35kV | 100 |
| | 133% | 5kV | 25 |
| | 133% | 15kV | 65 |
| | 133% | 25kV | 100 |

NOTE: For acceptance testing new cable only. Do not use these voltages for service-aged cable.
Derived from ANSI/IEEE Std. 141–1986 Table 82.

kilovolt. For example, if the maximum test voltage is 40kV and the number of steps is eight, the step voltage will be 40kV/8 = 5kV.

8. Make certain that all personnel are clear at *both* ends of the cable.

9. Energize the test set and *gradually* increase the voltage to the first step level (5kV from the example in step 7). Note that the current rises initially and gradually decreases. Wait one to five minutes to read and record the leakage current.

10. Gradually increase the voltage to the next step level (10kV in our example). When the next step voltage is reached, wait the same amount of time as in step 9. Read and record the leakage current.

11. Repeat step 10 for each of the voltage steps up to and including the maximum test voltage. **At any step, if the current continues to increase after the step voltage is reached, terminate the test immediately. The cable has failed.**

12. Leave the final voltage on the cable for fifteen minutes, reading and

**Table 4–3   Dielectric strength test record.**

CABLE _____ LOCATION _____

TYPE _____ RATING _____

DATE _____ WEATHER _____

REMARKS _____

| Af | | Bf | | Cf | |
|---|---|---|---|---|---|
| KV | mA | KV | mA | KV | mA |
| 5 | .8 | 5 | .4 | 5 | .6 |
| 10 | 1.2 | 10 | .8 | 10 | 1 |
| 15 | 1.8 | 15 | 1.25 | 15 | 1.4 |
| 20 | 2.6 | 20 | 1.6 | 20 | 2 |
| 25 | 3.2 | 25 | 2.15 | 25 | 2.6 |
| 30 | 3.95 | 30 | 2.6 | 30 | 3 |
| 35 | 4.6 | 35 | 3.15 | 35 | 3.6 |
| 1 MIN | | 1 MIN | | 1 MIN | 6.0 |
| 2 MIN | | 2 MIN | | 2 MIN | 4.75 |
| 3 MIN | | 3 MIN | | 3 MIN | 4.25 |
| 4 MIN | | 4 MIN | | 4 MIN | 3.80 |
| 5 MIN | | 5 MIN | | 5 MIN | 3.60 |
| 6 MIN | | 6 MIN | | 6 MIN | 3.58 |
| 7 MIN | | 7 MIN | | 7 MIN | 3.57 |
| 8 MIN | | 8 MIN | | 8 MIN | 3.55 |
| 9 MIN | | 9 MIN | | 9 MIN | 3.50 |
| 10 MIN | | 10 MIN | | 10 MIN | 3.48 |
| 11 MIN | | 11 MIN | | 11 MIN | 3.44 |
| 12 MIN | | 12 MIN | | 12 MIN | 3.42 |
| 13 MIN | | 13 MIN | | 13 MIN | 3.40 |
| 14 MIN | | 14 MIN | | 14 MIN | 3.40 |
| 15 MIN | | 15 MIN | | 15 MIN | 3.40 |

recording the leakage current every minute. The current should continue to decrease during this time period. All readings should be entered into a data table similar to the one shown in Table 4–3.

13. Gradually decrease the test voltage to zero. When the voltmeter on the test set indicates zero volts, put a ground wire on the cable and leave it for a minimum of 30 minutes.

## CAUTION:

THE CABLE WILL CONTINUE TO DRAIN CHARGE FOR THE ENTIRE PERIOD THAT THE GROUND WIRE IS ATTACHED. DO NOT REMOVE THE GROUND WIRE OR TOUCH THE CABLE WITHOUT WEARING PROPER INSULATING GLOVES AT ANY TIME!

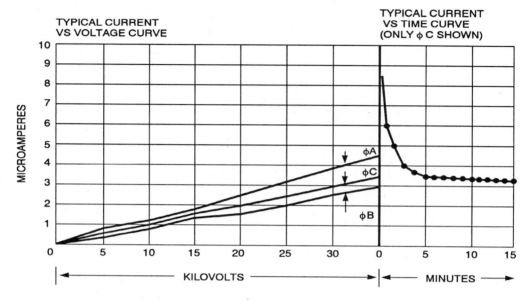

**Figure 4–13   Typical graph showing current vs voltage curves,
and current vs time curves.**

14. Plot the values of voltage versus current, and current versus time on a graph similar to the one shown in Figure 4-13.

## 0.1 Hertz Testing

As explained previously, the 0.1 Hertz test is becoming increasingly popular in some parts of the world. It offers the advantage of lighter equipment than 60 Hertz alternating voltage, yet it stresses the insulation as an alternating voltage would. Direct voltage testing can cause very large charge densities to build up in the area of insulation imperfections. These charge densities may take hours to drain even with a direct short across the cable insulation. If the cable is re-energized in such a way that the applied alternating voltage adds to the test charges, the cable may fail. This is not a problem with 0.1 Hertz testing, since the voltage alternates; therefore, no charge builds up, and the cable is not subjected to very high voltages when it is re-energized.

The problems described above are more of a concern during maintenance testing; new cable should not exhibit problems of this type.

## Power Factor Testing

The principles of power factor testing were explained earlier in this chapter. When a power factor test is performed on insulation, several quantities may be measured as follows.

- The watts consumed by the insulation may be measured. This is a measure of the absolute value of the leakage current; high values may indicate a general failure or deterioration of the insulation.

- The power factor (watts divided by volt-amperes) may be measured. This value can be compared to readings taken previously and from similar equipment. Greater than normal values usually indicate a problem.

- Power factor at different voltage levels can be measured. If the power factor increases as the voltage increases, the insulation may be contaminated.

Measurement of power factor on long cable runs is difficult because of the extremely high capacitive currents. These currents can be offset by the use of a parallel, resonating inductor. The inductor cancels out some of the capacitance of the cable. The readings are then taken as usual and compared to similar readings.

For the most part, power factor tests are usually not performed on cables.

## SAFETY PRECAUTIONS

**NOTE:**

The following safety precautions are typical and represent generally accepted safety practices. These precautions may not cover every installation or every variation. The worker should always refer to local safety standards, which always take precedence.

### Area Security

The immediate area at each end of the cable to be tested should be secured by constructing barricades and posting danger signs. A guard should be stationed at the remote end of the cable being tested and appropriate communications established.

### Equipment Isolation

All connected equipment should be disconnected from the cable before the test begins. If this is not possible, the test voltages should be reduced to levels that are acceptable for the connected equipment.

## Circuit and Cable Isolation

1. Identify a lead person for the job and assign duties.
2. Isolate, lock, tag, and test the equipment that is located in the testing area.
3. Post danger signs to identify the testing area.
4. Notify all personnel of the potentially hazardous conditions present.
5. Inspect all rubber protection equipment before it is placed in service.

## Safety Precautions Before Testing

1  Identify the tests to be performed and the test equipment to be used.
2. Identify the proper protective equipment for each test.
3. Review prints and clearances for each test.
4. Put rubber mats or sleeves on exposed conductors.
5. Be sure all switches involved in the circuits are opened.
6. Remove jumpers, potential transformers, lightning arresters, and other dangerous materials.
7. Be sure the cable under test is de-energized before touching it by testing the circuit with the proper voltage test instruments and performing the proper lockout/tagout procedures.
8. Ground the cable after the circuit is de-energized.

## Safety Precautions During Testing

1. Position personnel at both ends of the cable.
2. Apply ionization control (protection) where required on cable ends. Ionization control may be in the form of corona balls or insulating covers for the exposed cable ends.
3. Always wear appropriate rubber insulating clothing such as gloves and sleeves.

## Safety Precautions After Testing

1. Ground the tested conductor for a period of time *at least* long as the test voltage was present. Longer grounding times are preferred if possible.
2. Measure the conductor voltage after removing the ground and before touching the conductor. Wear rubber insulating equipment and always treat the exposed conductor as though it is still energized.

## Safety Grounding

In addition to grounding the cable for the purpose of draining the charge, cable should be routinely safety grounded at all times while personnel are working in the area. The following steps are recommended.

1. Always wear rubber gloves with appropriate leather protectors.
2. Verify that the circuit to be tested is de-energized. This is accomplished with the use of an appropriate voltage measuring tool. When measuring for a de-energized circuit, always check the instrument on a known energized source before and after measuring the circuit to be tested.
3. After the circuit is verified de-energized, an approved safety ground is first connected to the ground bus.
4. Connect the other end of the safety ground to the conductor. Make these connections with "hot sticks" when possible.
5. After each of the conductors are properly grounded, they can be disconnected and separated for testing.
6. The conductors should remain grounded at all times except when an individual conductor is under test. The safety grounds should first be removed from the cable, then from the ground.

# Chapter 5

# Cable Fault Location

## KEY POINTS

- What is the purpose of fault location?
- What are the common types of fault location equipment?
- What are the common methods of fault location?

## INTRODUCTION

Occasionally even newly installed cable will fail. When this happens, the cable may be completely replaced or repaired, depending on economics. If the decision is to repair the cable, the failed section of the cable must first be located.

This chapter describes the two most commonly employed methods of cable fault location and the equipment that is used.

## TERMS

The following terms and phrases are used in this chapter.

**Arc-reflecting:** A test set application wherein a test signal is bounced or reflected from a fault or short circuit.

**Characteristic Impedance:** The AC resistance that a cable exhibits to the passage of a pulse.

**CRO:** Abbreviation for Cathode Ray Oscilloscope. A CRO is a vacuum device used to display electrical signals or other information. The screen of a television set is a CRO.

**Fault:** A short circuit or other flaw in a power system conductor.

**TDR:** Time Domain Reflectometer (TDR) is an instrument that applies a low-energy voltage pulse to a cable. The TDR then displays the waves that are reflected from various faults and imperfections in the cable.

**Thumper:** A test instrument that applies a high energy voltage pulse to a cable that has failed.

## PURPOSE OF FAULT LOCATION

Power cables are subject to failure. A common practice has been to allow up to three splices in a cable before replacement. Before a cable can be repaired, the failed section must be accurately located. When the cable is above ground, as in cable tray installation, the location of the short circuit may be relatively easy. Buried cable or cable installed in conduit is a different matter.

A fault location technique needs to satisfy the following requirements.

- The fault must be located within a close enough tolerance that it can be found and repaired with a minimum amount of digging and confusion.
- The method used should be as nondestructive as possible. Minimum additional damage should be inflicted on the cable during the fault location process.

## METHODS

### Thumping

Thumping is the method of choice for locating high resistance faults. The basic principle in thumping involves impressing a repetitive, high energy pulse to the faulted line. The pulse is generated by rapidly discharging a capacitor to the cable. Because of the nature of the circuits involved, voltage peaks of as much as four times the thumper output voltage will be generated by the pulses that travel down the cable.

When a pulse reaches the fault, the excessive voltage will cause the faulted area to break down and arc. The fault may then be located in one of several ways.

- If the cable is direct buried, the fault can often be located by simply walking along the cable route until the vibration of the arc is felt in the ground.
- Audio sensing equipment may be used if the arc is not of sufficient strength to be felt.

There are several advantages and disadvantages to thumping. The advantages include:

- Thumping is relatively simple and straightforward. Little specialized technical skill is needed to locate the fault.
- Thumping will locate even very high resistance faults.

The disadvantages include:

- Since adjacent cables may be damaged by the thump, thumping may not be advisable for cable in cable trays or in multiple cable conduits.
- Some faults are very stubborn and may require hundreds or thousands of thumps to break the fault down.
- Because of the very high voltages and very rapid rise times, cable insulation is stressed to the maximum during the thumping process. If the thumping action takes too long, good insulation may be compromised.
- Acoustical echoes may make the location difficult to discern even when the fault breaks down.

In spite of the disadvantages, thumping is still the most popular of all the fault location methods.

## TDR

Power cable falls into the general category of a transmission line. At very high frequencies, all transmission lines possess an electrical property known as the characteristic impedance. The characteristic impedance of a given cable is a function of the cable material and dimensions. If the material or the dimensions change, the characteristic impedance will change. A splice, fault, bend, transformer, or other discontinuity in a cable will have a different characteristic impedance than the cable itself.

Time domain reflectometry takes advantage of the fact that a voltage pulse will create electrical echoes every time it passes an area with a different characteristic impedance. In this method, a low-energy, high-frequency pulse is applied to the cable. When the pulse reaches a point of irregularity, all or part of the energy is reflected back towards the input end of the cable.

At the input end of the cable, the time domain reflectometer uses a cathode ray

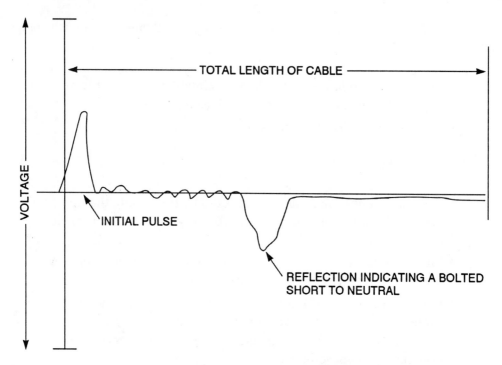

**Figure 5–1   Typical layout for TDR display.**
*(Courtesy of Cadick Professional Services)*

oscilloscope (CRO) to display the voltages that it detects. Figure 5–1 shows the general appearance of such a display.

The horizontal sweep control is adjusted so that the entire length of the cable is displayed on the CRO. The vertical sensitivity is adjusted so that the voltage pulses show clearly on the screen.

The initial pulse shows on the screen at the origin. The return pulse, if any, will reach the instrument at the same moment that the horizontal sweep reaches the point on the screen that represents the point from which the return pulse originated.

Different types of irregularities will produce differently shaped return pulses. Figure 5-2 shows examples of six different traces caused by various types of problems. This partial list represents the most common problems encountered.

The advantages of TDR include:

- TDR techniques are nondestructive.

- Generally provides a much more precise location of the fault.

- May allow the user to determine the type and degree of short circuit, thus allowing better splice planning.

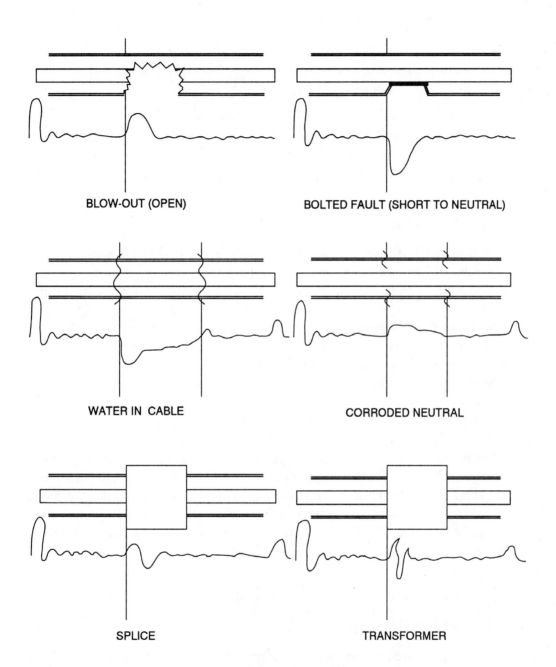

**Figure 5–2    Typical signatures for power utility applications.**
*(Courtesy of AVO Biddle® Instruments)*

The disadvantages of TDR include:

- TDR is extremely sensitive and picks up even minor problems. May make fault location difficult.
- Does not readily recognize high resistance short circuits.
- The CRO may be difficult to read, and generally requires an experienced operator.

## Arc Reflection Methods

At least one manufacturer has introduced a test method that operates using a combination of both thumping and time domain reflectometry. This system will operate as a TDR for all of the types of faults that the TDR can readily locate. For high-resistance short circuits, the operator can select a mode called arc reflection. In this mode, the test equipment applies a voltage pulse to the cable in much the same way that a thumper does, but the arc reflection system uses only the minimum amount of energy required to break the fault down into an arc.

At the same time that the arc creating voltage pulse is applied, the test equipment applies a high-frequency, low-energy pulse. The low energy pulse reflects off of the arc, and is shown on the CRO in much the same way were there a low-resistance fault present.

The major disadvantage of the TDR system is its inability to locate high-resistance faults. The arc reflection method eliminates this problem by using a voltage pulse to break down the fault. The major disadvantage to thumping is the likelihood that the cable may be damaged by the high-energy pulses. The arc reflection method minimizes this problem by using only the minimum amount of energy required to create the arc.

## TEST EQUIPMENT

Figures 5–3 and 5–4 show examples of typical TDRs. The hand-held model is a convenient portable package that is used for lighter duty applications. The larger model is a more deluxe model with built-in memory and a full range of calibration and adjustment features.

Figure 5–5 depicts a thumper. This unit applies a high-energy pulse to the cable and causes it to break down.

**Figure 5–3    Hand-held TDR.** *(Courtesy of AVO Biddle® Instruments)*

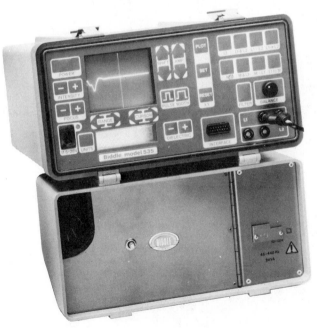

**Figure 5–4    Typical TDR unit.**
*(Courtesy of AVO Biddle® Instruments)*

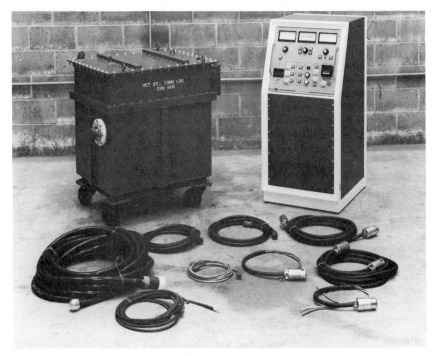

**Figure 5–5    Typical thumper on unit.**
*(Courtesy of AVO Biddle® Instruments)*

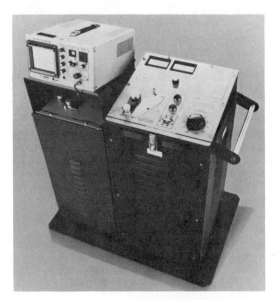

**Figure 5–6    Arc reflection cable fault locator.**
*(Courtesy of AVO Biddle® Instruments)*

Figure 5-6 shows an arc reflection unit. It may be considered to be a thumper in the same package with a TDR. The energy pulse that is produced is high enough to cause a low resistance arc to form, yet low enough to minimize or avoid any additional damage to the cable. Simultaneously, a low-energy, high-frequency pulse is placed on the cable. This pulse reflects off of the arc and puts a "blip" on the screen.

# PART TWO

# Chapter 6

---

# Cable Application
# Guide

---

## INTRODUCTION

One of the most often encountered problems in the field is when, where, and how to use the different types of cables and cable insulations available. This chapter is designed as an application guide, and includes information concerning the proper way to use these various cable and insulation types. This chapter includes information about UL-listed cable types and their insulation types.

UL-listed cables are assemblies that are manufactured for a specific application, and include types AC, FCC, IGS, MC, MI, MTW, MV, NM, NMC, SE, USE, TC, and UF.

Insulation materials used for the various UL-listed cable types are found in Table 310–13 of the NEC®, and include types FEP, FEPB, PF, PFAH, RH, RHH, RHW, RHW-2, SA, SIS, TA, TBS, TFE, THHN, THHW, THW, THWN, TW, V, XHHW, XHHW-2, Z, AND ZW.

### NOTE:

All of the cable types listed in this chapter are available from manufacturers. Some, such as varnished cambric (V), are not specified for new installations, and may be available only from specialty manufacturers. Cable types that previously used asbestos, such as FEBP and SA, are being replaced with other materials such as glass fiber.

The splicing and termination of these cable types is covered in Part One of this handbook. While special requirements have been noted in many cases, the user must refer to the manufacturer's recommendations and applicable *NEC®* sections for detailed information and regulations.

# ARMORED CABLE (TYPE AC)

## Description

**General.** Type AC cable is a fabricated assembly of insulated conductors in a flexible metal enclosure. Figure 6-1 shows a type of armored cable.

**Types.** Armored cable is available in the following types.

**AC:** No suffix indicates 60 degrees Celsius insulation. Conductors have thermoplastic insulation.
**ACH:** Suffix indicates 75 degrees Celsius insulation.
**ACHH:** Suffix indicates 90 degrees Celsius insulation.
**ACL:** Suffix indicates lead covering.
**ACT:** Suffix indicates thermosetting insulation.

## Applications

Armored cable is used in applications where the cable might be exposed to damaging conditions which cannot be minimized or eliminated by other types of systems.

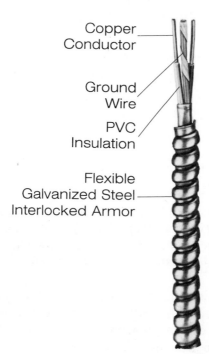

Copper Conductor

Ground Wire

PVC Insulation

Flexible Galvanized Steel Interlocked Armor

**Figure 6-1   Armored cable—Type AC.** *(Courtesy of General Cable Corporation)*

| **Permitted** | **Not Permitted** |
|---|---|
| Dry locations (Type AC) | Wet locations (Type AC) |
| Under plaster extensions | Theaters (518) |
| Embedded in plaster finish | Motion picture studios |
| On brick or other masonry | Hazardous locations (501 and 504) |
| | Corrosive fumes or vapors |
| Wet locations (Type ACL only) | Cranes or hoists (610) |
| | Storage battery rooms |
| | Hoistways or on elevators (620) |
| | Commercial garages (511) |

## NOTE:

Numbers in parentheses indicate articles in the *NEC®* where additional or explanatory information may be found.

## Sizes

| **Copper** | **Aluminum or Copper Coated Aluminum** |
|---|---|
| #14 AWG up to #1 AWG | #12 AWG up to #1 AWG |

## Temperature Ranges

Armored cable is available in temperature ranges **up to 90 degrees Celsius**. The allowed application temperature is determined by the type of insulation used on the conductors.

## Voltage Ranges

Armored cable is intended for use in **less than 2000 Volt** applications.

## Ampacities

Armored cable ampacities are determined using standard methods from the *NEC®*.

## Handling and Care

**Receiving and Handling.**
- Examine the protective covering for evidence of damage.

- Do not allow equipment to touch or damage cable surface or protective wrap.
- If a forklift is used, the forks must be long enough to contact both reel flanges.
- If a crane is used, a shaft through the arbor hold or flange cradles must be employed for the lift.
- If an inclined ramp is used for unloading, it must be wide enough to contact both flanges completely. The reel should be stopped at the bottom by using the flanges, not the cable surface.
- Never drop reels on the ground.

**Storage.**

- Unjacketed armored cable should be stored indoors to avoid corrosion.
- Reels should be stored on a hard clean surface, not bare earth.
- Store away from open fires or heat sources.
- Cable ends should be resealed as soon as a cable section has been removed.

**Receiving Tests.** Cable should be tested according to the insulation type and the methods outlined in Chapter 4 of this handbook.

## Installation

**Securing and Supporting.** Armored cable shall be secured and supported as follows.

- Approved staples, straps, hangers, or similar fittings as shown in Figure 6–2.

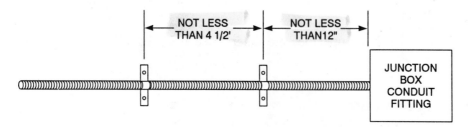

**Figure 6–2   Securing armored cable.**
*(Courtesy of Cadick Professional Services)*

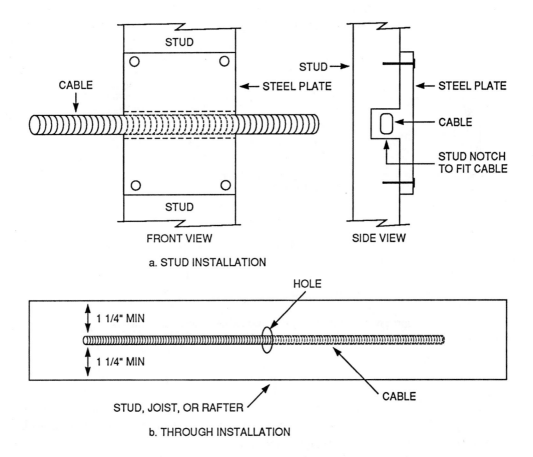

**Figure 6–3   Installing cable on studs, joists, and rafters.**
*(Courtesy of Cadick Professional Services)*

- When mounted on studs, joists, and rafters, type AC cable shall be installed as shown in Figure 6-3.

- When passed through bored holes, the edge of the hole shall be no less than 1 1/4 inches from the edge of the stud, joist, or rafter.

- In the event that 1 1/4 inches cannot be maintained, as mentioned previously and shown in Figure 6-3, a steel spacer or sleeve must be used to protect the cable.

- Armored cable installed in accessible attics shall be secured as shown in Figure 6-4.

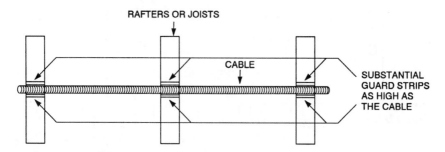

RAFTERS OR JOISTS

CABLE

SUBSTANTIAL
GUARD STRIPS
AS HIGH AS
THE CABLE

**a. OVER JOISTS OR RAFTERS IN ATTICS ACCESSIBLE BY LADDER OR STAIRS,
AS WITHIN SIX (6) FEET OF SCUTTLE HOLE OR ATTIC ENTRANCE**

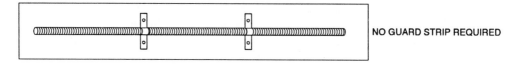

NO GUARD STRIP REQUIRED

**b. ALONG THE SIDE OF JOISTS, STUDS, OR RAFTER**

**Figure 6–4    Installing armored cable in accessible attics.**
*(Courtesy of Cadick Professional Services)*

**Termination and Splicing.** When armored cable is terminated, a fitting must be used to protect the insulation from abrasion, and an insulating bushing must be provided between the conductors and the armor.

If the armor is removed for the purpose of a splice or termination, the splice or termination must be made in a box or fitting approved for that purpose.

# FLUORINATED ETHYLENE PROPYLENE (TYPE FEP AND FEPB)

## Description

**General.** Type FEP and FEPB is an insulation type that is used in high temperature applications.

**Types.** Fluorinated ethylene propylene cable is available in the following types:

FEP        Dry and damp locations up to 90 degrees Celsius.
FEPB       Dry locations up to 200 degrees Celsius; special applications.

## Applications

**FEP**
Dry and damp locations up to 90 degrees Celsius

**FEPB**
Dry locations up to 200 degrees Celsius; special applications only

## Sizes

**FEP**  From **#14 AWG up to #2 AWG**.

**FEPB**  From **#14 AWG up to #8 AWG** with glass braid.
From **#6 AWG up to #2 AWG** with asbestos or other suitable material.

## Temperature Ranges

**Up to 200 degrees Celsius;** see applications table.

## Voltage Ranges

Circuits **below 600 Volts**.

## Ampacities

## Handling and Care

**Receiving and Handling.**

- Examine the protective covering for evidence of damage.
- Do not allow equipment to touch or damage cable surface or protective wrap.
- If a forklift is used, the forks must be long enough to contact both reel flanges.
- If a crane is used, a shaft through the arbor hold or flange cradles must be employed for the lift.
- If an inclined ramp is used for unloading, it must be wide enough to contact both flanges completely. The reel should be stopped at the bottom by using the flanges, not the cable surface.
- Never drop reels on the ground.

**Storage.**

- Reels should be stored on a hard clean surface, not bare earth.
- Store away from open fires or heat sources.

- Cable ends should be resealed as soon as a cable section has been removed.

**Receiving Tests.** Cable should be tested according to the insulation type and the methods outlined in Chapter 4 of this handbook.

## Installation

**Securing and Supporting.** Cable shall be secured and supported as follows.

- Approved staples, straps, hangers or similar fittings as shown in Figure 6-21.
- When mounted on studs, joists, and rafters, cable shall be installed as shown in Figure 6-22.
- When passed through bored holes, the edge of the hole shall be no less than 1 1/4 inches from the edge of the stud, joist, or rafter.
- In the event that 1 1/4 inches cannot be maintained, as mentioned previously and shown in Figure 6-3, a steel spacer or sleeve must be used to protect the cable.
- Cable installed in accessible attics shall be secured, as shown in Figure 6-23.

**Termination and Splicing.** Figure 6-24 shows the correct methods for termination of Fluorinated Ethylene Propylene cable.

## FLAT CONDUCTOR CABLE (TYPE FCC)

### Description

**General.** Type FCC cable consists of three or more flat copper conductors placed edge-to-edge, and separated and enclosed within an insulating assembly. Figure 6-5 shows a cross section of a typical flat conductor cable.

An FCC system is intended for use in branch circuits and is to be installed below carpet squares. It is installed as a complete system, including terminal boxes, receptacles, and wire; see Figure 6-6.

**Types.** N/A

INSTALLATION SYSTEM

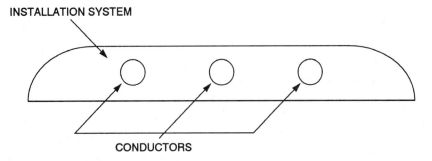

CONDUCTORS

**Figure 6–5   Cross–section of typical flat conductor cable.**
*(Courtesy of Cadick Professional Services)*

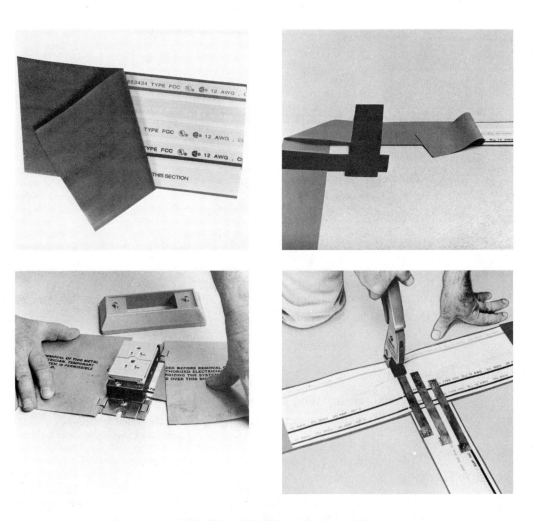

**Figure 6–6   Elements of a flat conductor cable system.**
*(Courtesy of AMP, Incorporated, Harrisburg, Pennsylvania)*

## Applications

FCC cable is intended for use in branch circuits to provide a complete and accessible power system. FCC cable allows for easy upgrading of existing office branch circuit power systems.

| **Permitted** | **Not Permitted** |
|---|---|
| Branch circuits: | Outdoors or in wet locations |
|    General-purpose | Where subject to corrosive vapors |
|    Appliance | Hazardous (Classified) locations |
|    Individual | Residential buildings |
| Floors: | School buildings |
|    Concrete | Hospital buildings |
|    Ceramic | |
|    Composition | |
|    Wood | |
| Walls: | |
|    In surface metal raceways | |
| Damp locations | |
| Heated floors | |
|    Must be made from materials | |
|    approved for floors above 30 | |
|    degrees Celsius | |

## Sizes

Three, four, and five conductors shall be arranged in each system. Wire sizes shall be of sufficient ampacity as described later.

## Temperature Ranges

FCC systems shall be installed at operating temperatures which are allowed according to the type of insulation; generally **60, 75, or 90 degrees Celsius.**

## Voltage Ranges

FCC cable shall be used in systems with **300 Volts maximum** between ungrounded conductors and **150 Volts maximum** between ungrounded and grounded conductors.

## Ampacities

General-purpose and appliance branch circuits:   **20 Amperes.**
Individual branch circuits:   **30 Amperes.**

## Handling and Care

### Receiving and Handling.

- Examine the protective covering for evidence of damage.
- Do not allow equipment to touch or damage cable surface or protective wrap.
- If a forklift is used, the forks must be long enough to contact both reel flanges.
- If a crane is used, a shaft through the arbor hold or flange cradles must be employed for the lift.
- If an inclined ramp is used for unloading, it must be wide enough to contact both flanges completely. The reel should be stopped at the bottom by using the flanges, not the cable surface.
- Never drop reels on the ground.

### Storage.

- Reels should be stored on a hard clean surface, not bare earth.
- Store away from open fires or heat sources.
- Cable ends should be resealed as soon as a cable section has been removed.

**Receiving Tests.** Cable should be tested according to the insulation type and the methods outlined in Chapter 4 of this handbook.

## Installation

**Securing and Supporting.** FCC systems cable shall be secured and supported as follows, and as shown in Figure 6-7.

- A metallic or nonmetallic bottom shield is mounted to the floor using an approved adhesive.
- The FCC cable and other components are placed on top of the shield.

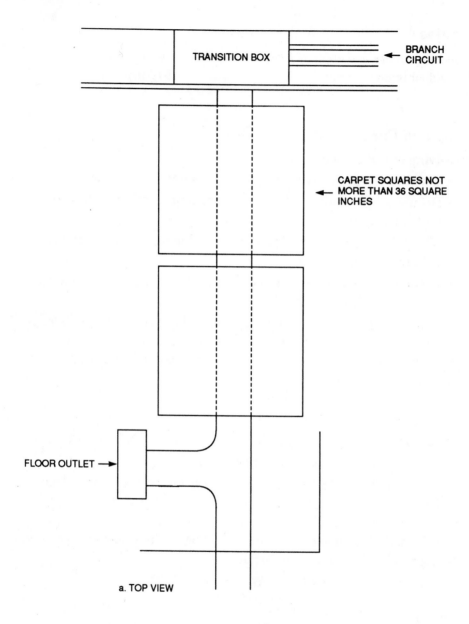

a. TOP VIEW

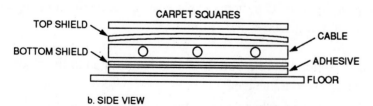

b. SIDE VIEW

**Figure 6–7    Installation of a flat conductor cable system.**
*(Courtesy of Cadick Professional Services)*

- A metallic top shield is placed over the FCC cable.
- The carpet square (36 square-inch maximum) is placed over the top shield.
- If any portion of the FCC system is more than 0.09 inch above the floor, it must be tapered or feathered.
- Crossing of more than two FCC cables at one point is not permitted.
- When any crossing occurs, the cables shall be separated by a grounded metal shield.

**Termination and Splicing.** All terminations and splices shall be made with equipment and materials specifically designed for FCC systems. Connections shall be installed such that electrical continuity, insulation, and sealing against dampness and liquid spillage are provided. Refer to the manufacturer's instructions for specific termination and splicing information.

## INTEGRATED GAS SPACER CABLE (TYPE IGS)

## Description

**General.** Figure 6–8 shows the general construction of type IGS cable. The following descriptions explain the basic construction.

> **Aluminum rods:** 250 kcmil (aluminum) conductors.
> **Strand shield:** Aluminum tape shield used to reduce voltage gradient stress and provide support for the conductors.
> **Paper spacer:** Provides support and insulation between the strand shield and the shield tape.
> **Shield tape:** Provides voltage gradient stress reduction and neutral paths.
> **$SF_6$ gas:** Insulation.
> **Conduit:** Mechanical protection and containment for the $SF_6$ gas.

**Types.** N/A

## Applications

Type IGS cable is used for underground feeder applications.

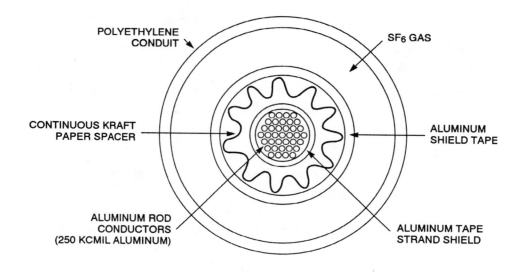

**Figure 6–8   Cross section of Type IGS cable.**
*(Courtesy of Cadick Professional Services)*

**Permitted**
Underground feeders:
   Direct burial
   Service entrance (USE)
   Feeders
   Branch circuits

**Not Permitted**
Interior wiring
Exposed in contact with buildings

## Sizes

250 kcmil up to 4750 kcmil.

## Temperature Ranges

N/A

## Voltage Ranges

0 Volts to 600 Volts.

## Ampacities

| kcmil | amperes |
|-------|---------|
| 250 | 119 |
| 500 | 168 |
| 750 | 206 |
| 1000 | 238 |
| 1250 | 266 |
| 1500 | 292 |
| 1750 | 315 |
| 2000 | 336 |
| 2250 | 357 |
| 2500 | 376 |
| 3000 | 412 |
| 3250 | 429 |
| 3500 | 445 |
| 3750 | 461 |
| 4000 | 476 |
| 4250 | 491 |
| 4500 | 505 |
| 4750 | 519 |

## Handling and Care

**Receiving and Handling.**

- Examine the protective covering for evidence of damage.
- Do not allow equipment to touch or damage cable surface or protective wrap.
- If a forklift is used, the forks must be long enough to contact both reel flanges.
- If a crane is used, a shaft through the arbor hold or flange cradles must be employed for the lift.
- If an inclined ramp is used for unloading, it must be wide enough to contact both flanges completely. The reel should be stopped at the bottom by using the flanges, not the cable surface.
- Never drop reels on the ground.

**Storage.**

- Reels should be stored on a hard clean surface, not bare earth.
- Store away from open fires or heat sources.
- Cable ends should be resealed as soon as a cable section has been removed.

**Receiving Tests.** Cable should be tested according to the insulation type and the methods outlined in Chapter 4 of this handbook.

## Installation

**Securing and Supporting.** IGS cable is intended for direct burial installation.

**Termination and Splicing.** Terminations and splices are required to be made with approved fittings suitable for maintaining the pressure of the $SF_6$ gas. A valve and cap is required for each length of cable. The valve and cap is used to measure gas pressure and to inject gas when required.

# METAL-CLAD CABLE (TYPE MC)

## Description

**General.** Type MC cable is a fabricated assembly of one or more individually insulated conductors in a metallic sheath. Figure 6-9 shows a type of metal-clad cable.

**Types.** Metal-clad cable is designated as type MC, with the following additions.

CS        Copper sheath.
ALS       Aluminum sheath.

## Applications

| Permitted | Not Permitted |
|---|---|
| Services, feeders, and branch circuits | Destructive corrosive conditions, |
| Power, lighting, control, and signal | except when the specific cable is |
| Indoors or outdoors | designed for such an application |
| Exposed or concealed | |
| Direct burial when so identified | |

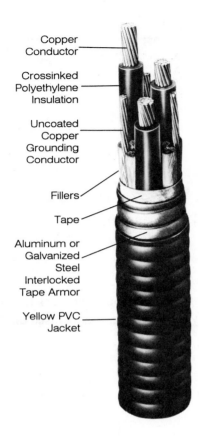

Copper Conductor

Crossinked Polyethylene Insulation

Uncoated Copper Grounding Conductor

Fillers

Tape

Aluminum or Galvanized Steel Interlocked Tape Armor

Yellow PVC Jacket

**Figure 6–9  Metal-clad cable.** *(Courtesy of General Cable Corporation)*

**Permitted**

Cable tray

Approved raceway

Open runs of cable

Aerial cable with messenger

Hazardous locations (501, 502, 503, 504)

Dry locations

Wet locations under any of the following conditions:

Metallic covering impervious to moisture

Lead sheath or other jacket

Insulated conductors suitable for wet applications

**Not Permitted**

**NOTE:**

Numbers in parentheses indicate articles in the *NEC*® where additional or explanatory information may be found.

## Sizes

**Copper**

**#18 AWG and larger**

**Aluminum or**
**Copper Coated Aluminum**
**#12 AWG and larger**

## Temperature Ranges

Metal-clad cable is available in temperature ranges **up to 90 degrees Celsius**. The allowed application temperature is determined by the type of insulation used on the conductors.

Insulations with higher allowable temperatures may be used in special applications.

## Voltage Ranges

Metal-clad cable is intended for use in circuits **up to 15,000 Volts**.

## Ampacities

Metal-clad cable ampacities are determined using standard methods from the *NEC*®.

## Handling and Care

**Receiving and Handling.**

- Examine the protective covering for evidence of damage.
- Do not allow equipment to touch or damage cable surface or protective wrap.
- If a forklift is used, the forks must be long enough to contact both reel flanges.
- If a crane is used, a shaft through the arbor hold or flange cradles must be employed for the lift.

- If an inclined ramp is used for unloading, it must be wide enough to contact both flanges completely. The reel should be stopped at the bottom by using the flanges, not the cable surface.
- Never drop reels on the ground.

**Storage.**
- Unjacketed, metal-clad cable should be stored indoors to avoid corrosion.
- Reels should be stored on a hard clean surface, not bare earth.
- Store away from open fires or heat sources.
- Cable ends should be resealed as soon as a cable section has been removed.

**Receiving Tests.** Cable should be tested according to the insulation type and the methods outlined in Chapter 4 of this handbook.

## Installation

**Securing and Supporting.** Metal-clad cable shall be secured and supported as follows.

- Approved staples, straps, hangers, or similar fittings as shown in Figure 6-10.
- In cable trays where approved for that application.
- Metal-clad cable may be direct buried as per Figure 6-11.
- Metal-clad cable may be used for overhead or underground service entrance cable.

**Termination and Splicing.** Termination of type MC cables should follow the procedures presented in Chapter 3. The metal sheath must be continuous and grounded.

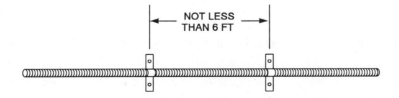

**Figure 6-10   Securing metal-clad cable.**
*(Courtesy of Cadick Professional Services)*

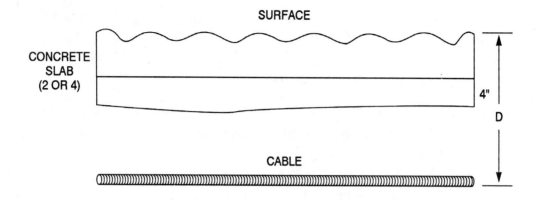

| LOCATION | D |
|---|---|
| 1) ALL LOCATIONS | 24" |
| 2) BELOW 2" CONCRETE PAD IN A TRENCH | 18" |
| 3) BENEATH A BUILDING (RACEWAY ONLY) | 0 |
| 4) UNDER 4" CONCRETE SLAB (NO TRAFFIC, SLAB OVER HANG 6" MIN.) | 18" |
| 5) STREET, HIGHWAYS, ETC. | 24" |
| 6) ONE AND TWO FAMILY DWELLING DRIVEWAYS | 18" |
| 7) AIRPORT RUNWAYS | 18" |

**Figure 6–11   Metal-clad direct burial cable.**
*(Courtesy of Cadick Professional Services)*

If a termination or splice is made at the junction with the cable system and a raceway system, it must be made in a box approved for that purpose.

## MINERAL INSULATED CABLE (TYPE MI)

### Description

**General.** Type MI cable consists of a copper conductor or conductors insulated with a highly compressed mineral insulation, usually magnesium oxide. The entire assembly is covered by a gastight and liquidtight copper or stainless steel sheath. Figure 6-12 shows the general construction of type MI cable.

**Types.** N/A

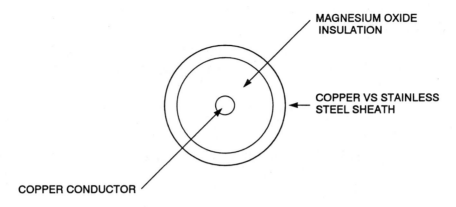

**Figure 6–12    Mineral insulated cable.**
*(Courtesy of Cadick Professional Services)*

## Applications

Type MI cable is used in applications where high resistance to fire damage is necessary. The metallic sheath and the magnesium oxide insulation are capable of withstanding fire temperatures in excess of one hour.

| Permitted | Not Permitted |
|---|---|
| Services, feeders, and branch circuits | Where exposed to damaging corrosion |
| Power, lighting, control, signal circuits | |
| Dry, wet, or continuously moist | |
| Indoors or outdoors | |
| Exposed or concealed | |
| Embedded in plaster, concrete or other | |
| Above or below grade | |
| Any hazardous location | |
| Exposed to oil or gasoline | |
| Corrosion not deteriorating to sheath | |
| Underground runs | |

## Sizes

UL listings provide for the following cable sizes.

| | |
|---|---|
| One conductor: | **#16 AWG up to 250 kcmil.** |
| Two & three conductors: | **#16 up to #4 AWG.** |
| Four conductors: | **#16 up to #6 AWG.** |
| Seven conductors: | **#16 up to #10 AWG.** |

Manufacturers can supply type MI cable in sizes **up to 500 kcmil.**

## Temperature Ranges

The operating temperature limits of MI type cable are determined by the fittings and terminations. They may be used **up to 85 degrees Celsius**.

## Voltage Ranges

MI cable is intended for use in **600 Volt** applications.

## Ampacities

| Conductors | Size | Amperes |
| --- | --- | --- |
| Single Conductors | 16 | 24 |
| | 14 | 35 |
| | 12 | 40 |
| | 10 | 55 |
| | 8 | 80 |
| | 6 | 105 |
| | 4 | 140 |
| | 3 | 165 |
| | 2 | 190 |
| | 1 | 220 |
| | 1/0 | 260 |
| | 2/0 | 300 |
| | 3/0 | 350 |
| | 4/0 | 405 |
| | 250 kcmil | 455 |
| Two Conductors | 16/2 | 16 |
| | 14/2 | 25 |
| | 12/2 | 32 |
| | 10/2 | 43 |
| | 8/2 | 59 |
| | 6/2 | 79 |
| | 4/2 | 104 |
| Three Conductors | 16/3 | 16 |
| | 14/3 | 25 |
| | 12/3 | 32 |

| | | |
|---|---|---|
| Three Conductors | 10/3 | 43 |
| (continued) | 8/3 | 59 |
| | 6/3 | 79 |
| | 4/3 | 104 |
| | | |
| Four Conductors | 16/4 | 12.8 |
| | 14/4 | 10.0 |
| | 12/4 | 25.6 |
| | 10/4 | 34.4 |
| | 8/4 | 47.2 |
| | 6/4 | 63.2 |
| | | |
| Seven Conductors | 16/7 | 11.2 |
| | 14/7 | 17.5 |
| | 12/7 | 22.4 |
| | 10/7 | 30.1 |

*(Courtesy of Pyrotenax Cable)*

## Handling and Care

### Receiving and Handling.

- Do not allow equipment to touch or damage cable surface or protective wrap.
- If a forklift is used, the forks must be long enough to contact both reel flanges.
- If a crane is used, a shaft through the arbor hold or flange cradles must be employed for the lift.
- If an inclined ramp is used for unloading, it must be wide enough to contact both flanges completely. The reel should be stopped at the bottom by using the flanges, not the cable surface.
- Never drop reels on the ground.

### Storage.

- Reels should be stored on a hard clean surface, not bare earth.
- Store away from open fires or heat sources.
- Cable ends should be resealed as soon as a cable section has been removed.

**Receiving Tests.** Cable should be tested according to the insulation type and the methods outlined in Chapter 4 of this handbook.

## Installation

**Securing and Supporting.** Type MI cable shall be secured and supported as follows.

- Approved staples, straps, hangers or similar fittings as shown in Figure 6-13.
- When mounted on studs, joists, and rafters, type AC cable shall be installed as shown in Figure 6-14.
- When passed through bored holes, the edge of the hole shall be no less than 1 1/4 inches from the edge of the stud, joist, or rafter.
- In the event that 1 1/4 inches cannot be maintained, as mentioned previously and shown in Figure 6-14, a steel spacer or sleeve must be used to protect the cable.
- Mineral insulated cable installed in accessible attics shall be secured, as shown in Figure 6-15.

**Termination and Splicing.** Figure 6-16 shows a typical termination for mineral insulated cable. Critical items in this termination include the following.

- Exposure of the mineral insulation to air should be minimized.
- The termination should completely seal the insulation from the outside.
- The copper shield should be used as a bonding conductor only when approved by the manufacturer for such service.

Type MI cable does not retain its fire temperature rating when field spliced. The only way to splice MI cable and retain its fire rating is with a factory splice. Note that:

- The splice is actually a double termination with a copper shield placed over them to provide mechanical strength.
- As previously mentioned, the field splice is not rated for the fire

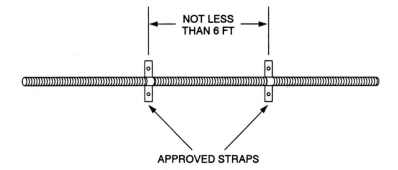

**Figure 6–13   Installation of Type MI cable.** *(Courtesy of Cadick Professional Services)*

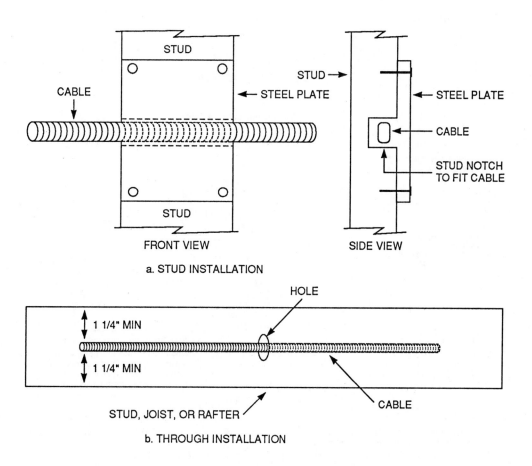

**Figure 6–14   Installing cable on studs, joists, and rafters.**
*(Courtesy of Cadick Professional Services)*

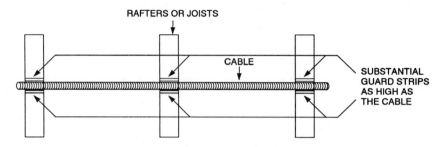

a. OVER JOISTS OR RAFTERS IN ATTICS ACCESSIBLE BY LADDER OR STAIRS,
AS WITHIN SIX (6) FEET OF SCUTTLE HOLE OR ATTIC ENTRANCE

NO GUARD STRIP REQUIRED

b. ALONG THE SIDE OF JOISTS, STUDS, OR RAFTER

**Figure 6–15    Installing cable in accessible attics.**
*(Courtesy of Cadick Professional Services)*

**Terminating 600 Volt Cable**

ONE "PYROPAK" CONTAINS SUFFICIENT MATERIAL TO TERMINATE TWO ENDS OF CABLE.

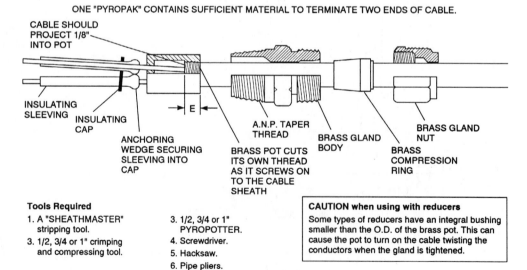

CABLE SHOULD
PROJECT 1/8"
INTO POT

INSULATING
SLEEVING

INSULATING
CAP

ANCHORING
WEDGE SECURING
SLEEVING INTO
CAP

E

A.N.P. TAPER
THREAD

BRASS POT CUTS
ITS OWN THREAD
AS IT SCREWS ON
TO THE CABLE
SHEATH

BRASS GLAND
BODY

BRASS
COMPRESSION
RING

BRASS GLAND
NUT

**Tools Required**

1. A "SHEATHMASTER"
   stripping tool.
3. 1/2, 3/4 or 1" crimping
   and compressing tool.

3. 1/2, 3/4 or 1"
   PYROPOTTER.
4. Screwdriver.
5. Hacksaw.
6. Pipe pliers.

> **CAUTION when using with reducers**
> Some types of reducers have an integral bushing
> smaller than the O.D. of the brass pot. This can
> cause the pot to turn on the cable twisting the
> conductors when the gland is tightened.

**CAUTION**

The magnesium oxide insulation in this cable will absorb moisture when exposed to air for any length of time. It is desirable therefore to complete a termination once started.

**Figure 6–16    Mineral insulated cable terminations.**
*(Courtesy of Pyrotenax USA, Inc., East Syracuse, NY)*

temperature rating of the cable itself. Field splices are typically rated at a maximum of 200 to 260 degrees Celsius.

# MACHINE TOOL WIRE (TYPE MTW)

## Description

**General.** Type MTW is made of a single copper conductor surrounded by thermoplastic insulation resistant to moisture, heat, and oil.

**Types.** N/A

## Applications

Machine tool wire is used in applications where the cable might be exposed to damaging conditions that cannot be minimized or eliminated by other types of systems.

## Sizes

Sizes are available from **#22 AWG up to 1000 kcmil.**

## Temperature Ranges

Type MTW is available in both **60 and 90 degrees Celsius** insulation.

## Voltage Ranges

Type MTW wire is intended for use in applications **600 Volts and below**.

## Ampacities

Type MTW wire ampacities are determined using standard methods from the *NEC*®.

## Handling and Care

**Receiving and Handling.**
- Examine the protective covering for evidence of damage.
- Do not allow equipment to touch or damage cable surface or protective wrap.

- If a forklift is used, the forks must be long enough to contact both reel flanges.
- If a crane is used, a shaft through the arbor hold or flange cradles must be employed for the lift.
- If an inclined ramp is used for unloading, it must be wide enough to contact both flanges completely. The reel should be stopped at the bottom by using the flanges, not the cable surface.
- Never drop reels on the ground.

**Storage.**

- Reels should be stored on a hard clean surface, not bare earth.
- Store away from open fires or heat sources.
- Cable ends should be resealed as soon as a cable section has been removed.

**Receiving Tests.** Cable should be tested according to the insulation type and the methods outlined in Chapter 4 of this handbook.

## Installation

**Securing and Supporting.** MTW wire should be installed and supported as described in NFPA Standard 79. Refer to the latest version of that standard for more specific information.

**Termination and Splicing.** Type MTW wire is terminated using standard methods, as shown in Figure 6–17. Splicing should be avoided with type MTW.

## MEDIUM-VOLTAGE CABLE (TYPE MV)

### Description

**General.** Type MV cable is a single or multi-conductor solid dielectric insulated cable rated for 2001 Volts and higher. Figure 6–18 shows a typical MV cable.

**Types.** N/A

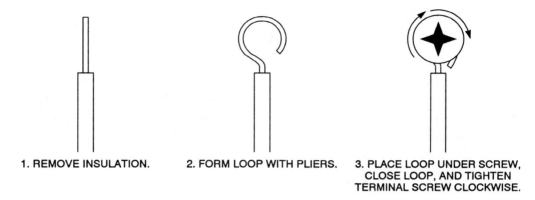

1. REMOVE INSULATION.    2. FORM LOOP WITH PLIERS.    3. PLACE LOOP UNDER SCREW,
                                                         CLOSE LOOP, AND TIGHTEN
                                                         TERMINAL SCREW CLOCKWISE.

**Figure 6–17   Terminating Type MTW cable.**

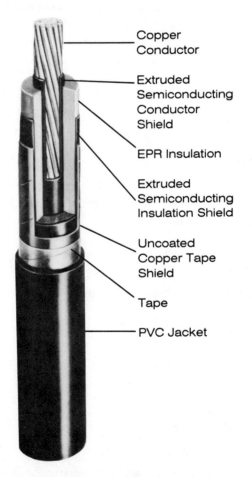

Copper
Conductor

Extruded
Semiconducting
Conductor
Shield

EPR Insulation

Extruded
Semiconducting
Insulation Shield

Uncoated
Copper Tape
Shield

Tape

PVC Jacket

**Figure 6–18   Type MV–90 cable.**
*(Courtesy of General Cable Corporation)*

## Applications

MV cables are used on systems up to 35,000 Volts as follows:

| **Permitted** | **Not Permitted** |
|---|---|
| Wet or dry locations | Direct sunlight, unless identified for |
| Raceways | Cable trays, unless identified for |
| Cable trays (318) | |
| Direct burial (710) | |
| Messenger supported | |

### NOTE:

Numbers in parentheses indicate articles in the *NEC*® where additional or explanatory information may be found.

## Sizes

Type MV cables are available in sizes **up to 4000 kcmil,** but common applications normally only require sizes up to 750 kcmil.

## Temperature Ranges

Type MV cable temperature ranges are determined by the type of insulation used. Standard values **up to 90 degrees Celsius** are used, however high temperature insulations are available.

## Voltage Ranges

Type MV cable is available from **2001 Volts up to 35,000 Volts.**

## Ampacities

Type MV cable ampacities are determined using standard methods from the *NEC*®.

## Handling and Care

### Receiving and Handling.

- Examine the protective covering for evidence of damage.
- Do not allow equipment to touch or damage cable surface or protective wrap.

- If a forklift is used, the forks must be long enough to contact both reel flanges.
- If a crane is used, a shaft through the arbor hold or flange cradles must be employed for the lift.
- If an inclined ramp is used for unloading, it must be wide enough to contact both flanges completely. The reel should be stopped at the bottom by using the flanges, not the cable surface.
- Never drop reels on the ground.

**Storage.**
- Reels should be stored on a hard clean surface, not bare earth.
- Store away from open fires or heat sources.
- Cable ends should be resealed as soon as a cable section has been removed.

**Receiving Tests.** Cable should be tested according to the insulation type and the methods outlined in Chapter 4 of this handbook.

## Installation

**Securing and Supporting.** MV cable may be installed above the ground as follows.

- Rigid metal conduit
- Intermediate metal conduit
- Rigid nonmetallic conduit
- Cable trays
- Busways and cable buses

MV cable may be direct buried when so identified, and as shown in Figure 6-19.

When direct buried, MV cable must be shielded, or be in a continuous metal sheath.

**Termination and Splicing.** Terminations and splicing of medium-voltage cable is covered in detail in Chapter 3.

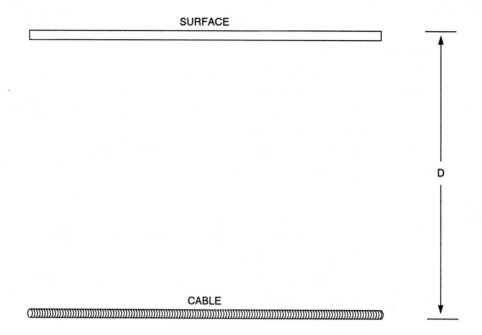

| VOLTAGE | D (MINIMUM) |
|---|---|
| BELOW 22KV | 30 INCHES |
| 22K - 40KV | 36 INCHES |
| OVER 40KV | 42 INCHES |

**Figure 6–19   MV direct burial cable.**
*(Courtesy of Cadick Professional Services)*

# NONMETALLIC-SHEATHED CABLE (TYPE NM AND NMC)

## Description

**General.** Nonmetallic-sheathed cable, shown in Figure 6–20, is a factory assembly of two or more insulated conductors having an outer sheath of moisture-resistant, flame-retardant, nonmetallic material.

**Types.** Nonmetallic-sheathed cable is available in the following types.

NM      Normally dry locations.
NMC     Corrosion resistant for dry, moist, damp, or corrosive llocations.
        This type of wire has been given the field name of "Romex."

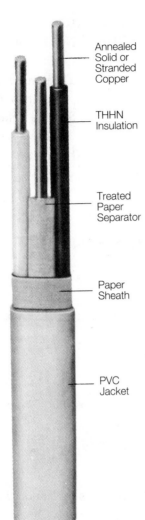

Annealed
Solid or
Stranded
Copper

THHN
Insulation

Treated
Paper
Separator

Paper
Sheath

PVC
Jacket

**Figure 6–20    Nonmetallic-sheathed cable.**
*(Courtesy of General Cable Corporation)*

## Applications

Permitted locations include the following:

| NM | | NMC | |
|---|---|---|---|
| **Permitted** | **Not Permitted** | **Permitted** | **Not Permitted** |
| Exposed | Dwelling above three | Exposed and | Dwellings above |
| Concealed | stories | concealed | three stories |
| Normally dry | Service entrance | Dry, moist, | Service entrance |
| locations | cable | damp, or | cable |
| | | corrosive | |
| | | locations | |

|  | NM |  | NMC |
|---|---|---|---|
| **Permitted** | **Not Permitted** | **Permitted** | **Not Permitted** |
| Air voids, when not moist | Commercial garages | Outside or inside walls | Commercial garages |
|  | Theaters or motion picture studios |  | Theaters or motion picture studios |
|  | Storage battery rooms |  | Storage battery rooms |
|  | Hoistways where exposed to corrosive fumes |  | Hoistways |

## Sizes

Common sizes available include **#6, #8, #10, #12, and #14 AWG.**

## Temperature Ranges

Application temperatures are determined by the insulation type and may be found in the appropriate tables in the *NEC®*.

## Voltage Ranges

Nonmetallic sheathed cable is for use in circuits **below 600 Volts.** Normal application is for 120/240 Volt residential service.

## Ampacities

Nonmetallic sheathed cable ampacities are determined using standard methods from the *NEC®*.

## Handling and Care

**Receiving and Handling.**

- Examine the protective covering for evidence of damage.
- Do not allow equipment to touch or damage cable surface or protective wrap.
- If a forklift is used, the forks must be long enough to contact both reel flanges.
- If a crane is used, a shaft through the arbor hold or flange cradles must be employed for the lift.
- If an inclined ramp is used for unloading, it must be wide

enough to contact both flanges completely. The reel should be stopped at the bottom by using the flanges, not the cable surface.

- Never drop reels on the ground.

**Storage.**

- Reels should be stored on a hard clean surface, not bare earth.
- Store away from open fires or heat sources.
- Cable ends should be resealed as soon as a cable section has been removed.

**Receiving Tests.** Cable should be tested according to the insulation type and the methods outlined in Chapter 4 of this handbook.

## Installation

**Securing and Supporting.** Nonmetallic sheathed cable shall be secured and supported as follows:

- Approved staples, straps, hangers or similar fittings, as shown in Figure 6–21.
- When mounted on studs, joists, and rafters, nonmetallic sheathed cable shall be installed as shown in Figure 6–22.
- When passed through bored holes, the edge of the hole shall be no less than 1 1/4 inches from the edge of the stud, joist, or rafter.

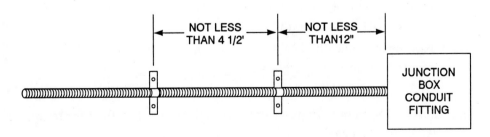

**Figure 6–21   Securing nonmetallic-sheathed cable.**
*(Courtesy of Cadick Professional Services)*

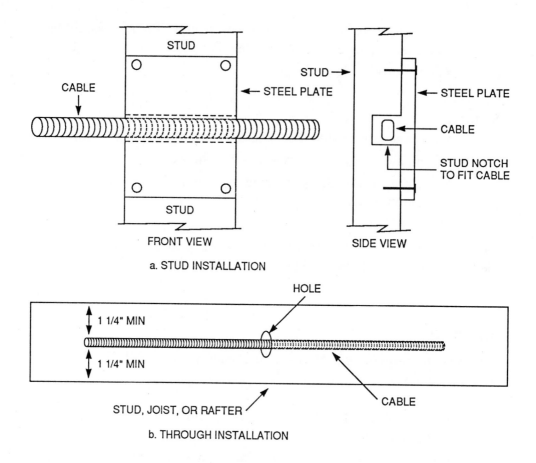

**Figure 6–22   Installing cable on studs, joists, and rafters.**
*(Courtesy of Cadick Professional Services)*

- In the event that 1 1/4 inches cannot be maintained, as mentioned previously and shown in Figure 6-3, a steel spacer or sleeve must be used to protect the cable.

- Nonmetallic sheathed cable installed in accessible attics shall be secured, as shown in Figure 6-23.

**Termination and Splicing.** Figure 6-24 shows the correct method for termination of nonmetallic sheathed cable.

Nonmetallic sheathed cable splices and terminations must be made in an electrical box approved for such service.

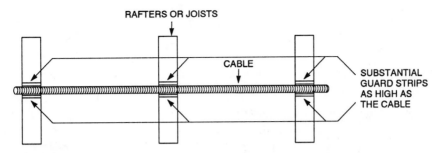

a. OVER JOISTS OR RAFTERS IN ATTICS ACCESSIBLE BY LADDER OR STAIRS,
AS WITHIN SIX (6) FEET OF SCUTTLE HOLE OR ATTIC ENTRANCE

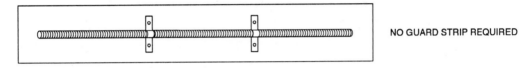

NO GUARD STRIP REQUIRED

b. ALONG THE SIDE OF JOISTS, STUDS, OR RAFTER

**Figure 6–23   Installing nonmetallic cable in accessible attics.**
*(Courtesy of Cadick Professional Services)*

# PERFLUOROALKOXY (TYPE PFA AND PFAH)

## Description

**General.**  Perfluoroalkoxy is an insulation type used for general wiring and for high temperature applications.

**Types.**  Perfluoroalkoxy cable is available in the following types.

| | |
|---|---|
| **PFA** | Dry locations up to 200 degrees Celsius. |
| | Dry and wet locations up to 90 degrees Celsius. |
| **PFAH** | Dry locations up to 250 degrees Celsius. |

## Applications

### PFA
Dry and damp locations up to 90 degrees Celsius
Dry locations up to 200 degrees Celsius

### PFAH
Dry locations up to 250 degrees Celsius.
Only for leads within apparatus or within raceways connected to apparatus.

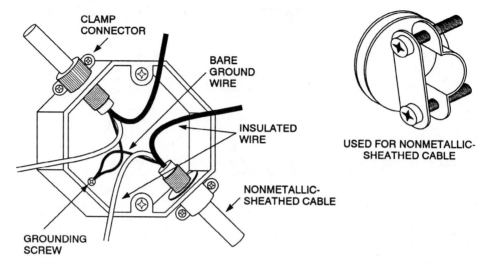

a. BOX TERMINATION OF CABLE

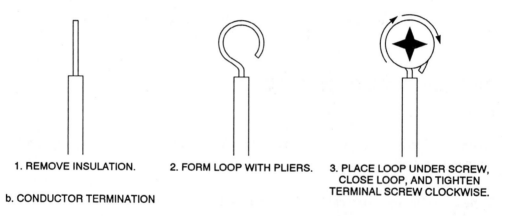

1. REMOVE INSULATION.    2. FORM LOOP WITH PLIERS.    3. PLACE LOOP UNDER SCREW, CLOSE LOOP, AND TIGHTEN TERMINAL SCREW CLOCKWISE.

b. CONDUCTOR TERMINATION

**Figure 6–24    Terminating nonmetallic-sheathed cables.**

## Sizes

From **#14 AWG up to 4/0 AWG.**

## Temperature Ranges

**Up to 250 degrees Celsius;** see applications table.

## Voltage Ranges

Circuits **below 600 Volts**. Normal application is for 120/240 Volt residential service.

## Ampacities

Ampacities are determined using standard methods from the *NEC*®.

## Handling and Care

### Receiving and Handling.

- Examine the protective covering for evidence of damage.
- Do not allow equipment to touch or damage cable surface or protective wrap.
- If a forklift is used, the forks must be long enough to contact both reel flanges.
- If a crane is used, a shaft through the arbor hold or flange cradles must be employed for the lift.
- If an inclined ramp is used for unloading, it must be wide enough to contact both flanges completely. The reel should be stopped at the bottom by using the flanges, not the cable surface.
- Never drop reels on the ground.

### Storage.

- Reels should be stored on a hard clean surface, not bare earth.
- Store away from open fires or heat sources.
- Cable ends should be resealed as soon as a cable section has been removed.

### Receiving Tests.
Cable should be tested according to the insulation type and the methods outlined in Chapter 4 of this handbook.

## Installation

### Securing and Supporting.
Cable shall be secured and supported as follows.

- Approved staples, straps, hangers or similar fittings, as shown in Figure 6-21.
- When mounted on studs, joists, and rafters, cable shall be installed as shown in Figure 6-22.
- When passed through bored holes, the edge of the hole shall be

no less than 1 1/4 inches from the edge of the stud, joist, or rafter.

- In the event that 1 1/4 inches cannot be maintained, as mentioned previously and shown in Figure 6-3, a steel spacer or sleeve must be used to protect the cable.
- Cable installed in accessible attics shall be secured as shown in Figure 6-23.

**Termination and Splicing.**  Figure 6-24 shows the correct method for termination of Perfluoroalkoxy cable.

# HEAT-RESISTANT RUBBER (TYPE RH, RHH, RHW, AND RHW-2)

## Description

**General.**  Heat-resistant rubber is an insulation type which is used for general wiring applications.

**Types.**  Heat resistant rubber cable is available in the following types:

| | |
|---|---|
| **RH** | Dry and damp locations up to 75 degrees Celsius. |
| **RHH** | Dry and damp locations up to 90 degrees Celsius. |
| **RHW** | Dry and wet locations up to 75 degrees Celsius. |
| **RHW-2** | Dry and wet locations up to 90 degrees Celsius. |

## Applications

Used for all general wiring requirements for low- and medium-voltage applications.

## Sizes

From **#14 AWG up to 2000 kcmil.**

## Temperature Ranges

**Up to 90 degrees Celsius.**

## Voltage Ranges

Circuits **below 600 Volts up to 35,000 Volts.**

## Ampacities

Ampacities are determined using standard methods from the *NEC®*.

## Handling and Care

**Receiving and Handling.**

- Examine the protective covering for evidence of damage.
- Do not allow equipment to touch or damage cable surface or protective wrap.
- If a forklift is used, the forks must be long enough to contact both reel flanges.
- If a crane is used, a shaft through the arbor hold or flange cradles must be employed for the lift.
- If an inclined ramp is used for unloading, it must be wide enough to contact both flanges completely. The reel should be stopped at the bottom by using the flanges, not the cable surface.
- Never drop reels on the ground.

**Storage.**

- Reels should be stored on a hard clean surface, not bare earth.
- Store away from open fires or heat sources.
- Cable ends should be resealed as soon as a cable section has been removed.

**Receiving Tests.** Cable should be tested according to the insulation type and the methods outlined in Chapter 4 of this handbook.

## Installation

**Securing and Supporting.** Low-voltage cable shall be secured and supported as follows.

- Approved staples, straps, hangers or similar fittings, as shown in Figure 6–21.

- When mounted on studs, joists, and rafters, cable shall be installed as shown in Figure 6-22.
- When passed through bored holes, the edge of the hole shall be no less than 1 1/4 inches from the edge of the stud, joist, or rafter.
- In the event that 1 1/4 inches cannot be maintained, as mentioned previously and shown in Figure 6-3, a steel spacer or sleeve must be used to protect the cable.
- Cable installed in accessible attics shall be secured as shown in Figure 6-23.

Medium-voltage cable should be installed as described previously for type MV cable.

**Termination and Splicing.** Terminations depend upon application voltages. See the appropriate sections for complete information.

## TYPE SA CABLE

### Description

**General.**   Type SA cable employs an insulation system which features a silicone insulation layer covered by a heat resistant aramid or glass jacket.  Type SA wire provides relatively high ampacities and good temperature performance; see Figure 6-25.

**Types.**  N/A

### Applications

Used for general wiring applications in high ambient temperature locations, including power plants, boilers, and ovens.

**Figure 6-25   Type SA 600V power and control cable.**
*(Courtesy of The Okonite Company—Power, Control, and Instrumentation Cables)*

## Sizes

From **#14 AWG up to 2000 kcmil.**

## Temperature Ranges

**Up to 90 degrees Celsius** in dry and damp locations.
**Up to 125 degrees Celsius** in special applications.

## Voltage Ranges

Circuits **below 600 Volts.**

## Ampacities

Ampacities are determined using standard methods from the *NEC*®.

## Handling and Care

**Receiving and Handling.**

- Examine the protective covering for evidence of damage.
- Do not allow equipment to touch or damage cable surface or protective wrap.
- If a forklift is used, the forks must be long enough to contact both reel flanges.
- If a crane is used, a shaft through the arbor hold or flange cradles must be employed for the lift.
- If an inclined ramp is used for unloading, it must be wide enough to contact both flanges completely. The reel should be stopped at the bottom by using the flanges, not the cable surface.
- Never drop reels on the ground.

**Storage.**

- Reels should be stored on a hard clean surface, not bare earth.
- Store away from open fires or heat sources.
- Cable ends should be resealed as soon as a cable section has been removed.

**Receiving Tests.** Cable should be tested according to the insulation type and the methods outlined in Chapter 4 of this handbook.

## Installation

**Securing and Supporting.** Cable shall be secured and supported as follows.

- Approved staples, straps, hangers or similar fittings, as shown in Figure 6-21.
- When mounted on studs, joists, and rafters, cable shall be installed as shown in Figure 6-22.
- When passed through bored holes, the edge of the hole shall be no less than 1 1/4 inches from the edge of the stud, joist, or rafter.
- In the event that 1 1/4 inches cannot be maintained, as mentioned previously and shown in Figure 6-3, a steel spacer or sleeve must be used to protect the cable.
- Cable installed in accessible attics shall be secured as shown in Figure 6-23.

**Termination and Splicing.** Terminations depend upon application voltages. See the appropriate sections for complete information.

# SERVICE ENTRANCE CABLE (TYPE SE AND USE)

## Description

**General.** Service entrance cable is a single conductor or multi-conductor assembly provided with or without an overall covering, and is primarily used in service entrances. Conductors used may be Type RH, RHW, RHH, or XHHW; see Figure 6-26.

**Types.** Service entrance cable is available in the following types:

| | |
|---|---|
| SE | Cable with a flame retardant, moisture resistant covering. |
| USE | Cable for underground service use. It has a moisture resistant covering and sometimes a flame-retardant covering also. |
| USE-2 | Same as type USE but with higher rated insulation. |

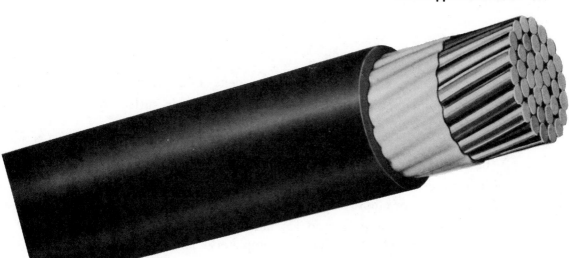

**Figure 6–26    USE–2 cable.** *(Courtesy of Rome Cable Corporation)*

## Applications

- Service entrance conductors; Article 230 of the *NEC*®.
- Branch circuits or feeders with insulated ground conductor.

## Sizes

Sizes **up to 2000 kcmil** are available.

## Temperature Ranges

Types SE and USE: **75 degrees Celsius.**
Type USE-2: **90 degrees Celsius.**

## Voltage Ranges

Intended for applications **600 Volts and below**.

## Ampacities

Service entrance cable ampacities are determined using standard methods from the *NEC*®.

# Handling and Care

## Receiving and Handling.

- Examine the protective covering for evidence of damage.
- Do not allow equipment to touch or damage cable surface or protective wrap.
- If a forklift is used, the forks must be long enough to contact both reel flanges.
- If a crane is used, a shaft through the arbor hold or flange cradles must be employed for the lift.
- If an inclined ramp is used for unloading, it must be wide enough to contact both flanges completely. The reel should be stopped at the bottom by using the flanges, not the cable surface.
- Never drop reels on the ground.

## Storage.

- Reels should be stored on a hard clean surface, not bare earth.
- Store away from open fires or heat sources.
- Cable ends should be resealed as soon as a cable section has been removed.

**Receiving Tests.** Cable should be tested according to the insulation type and the methods outlined in Chapter 4 of this handbook.

# Installation

**Securing and Supporting.** When used in interior wiring, service entrance cable shall be secured and supported as follows.

- Approved staples, straps, hangers or similar fittings, as shown in Figure 6-2.
- When mounted on studs, joists, and rafters, type SE and USE cable shall be installed, as shown in Figure 6-3.
- When passed through bored holes, the edge of the hole shall be no less than 1 1/4 inches from the edge of the stud, joist, or rafter.

- In the event that 1 1/4 inches cannot be maintained, as mentioned previously and shown in Figure 6-3, a steel spacer or sleeve must be used to protect the cable.
- Service entrance cable installed in accessible attics shall be secured as shown in Figure 6-4.

When installed as service entrance, cable installation shall comply with Article 230 of the *NEC*®.

**Termination and Splicing.** Termination of service entrance cable shall be made according to the procedures discussed in Chapter 3. Splices and termination shall be made in a box or fitting approved for such service.

## SYNTHETIC HEAT RESISTANT (TYPE SIS)

### Description

**General.** Synthetic heat-resistant wire (type SIS) is a single conductor copper wire insulated with heat resistant rubber. It is intended for use in wiring switchboards and other control panel applications. The conductor of type SIS may be solid or stranded.

**Types.** N/A

### Applications

For switchboard wiring only.

### Sizes

From **#14 AWG up to 4/0 AWG.**

### Temperature Ranges

**Up to 90 degrees Celsius.**

## Voltage Ranges

Circuits **below 600 Volts**. Normal application is in low voltage control applications.

## Ampacities

Ampacities are determined using standard methods from the NEC®.

## Handling and Care

**Receiving and Handling.**

- Examine the protective covering for evidence of damage.
- Do not allow equipment to touch or damage cable surface or protective wrap.
- If a forklift is used, the forks must be long enough to contact both reel flanges.
- If a crane is used, a shaft through the arbor hold or flange cradles must be employed for the lift.
- If an inclined ramp is used for unloading, it must be wide enough to contact both flanges completely. The reel should be stopped at the bottom by using the flanges, not the cable surface.
- Never drop reels on the ground.

**Storage.**

- Reels should be stored on a hard clean surface, not bare earth.
- Store away from open fires or heat sources.
- Cable ends should be resealed as soon as a cable section has been removed.

**Receiving Tests.** Cable should be tested according to the insulation type and the methods outlined in Chapter 4 of this handbook.

## Installation

**Securing and Supporting.** The installation of type SIS is determined by the particular switchboard application. Cable clamps, cable ties, and lacing twine are all acceptable methods.

**Termination and Splicing.** Type SIS may be terminated using squeeze-on lugs. Smaller sizes may be terminated directly using methods shown in Figure 6–17.

## THERMOPLASTIC ASBESTOS (TYPE TA)

### Description

**General.** Thermoplastic asbestos is an insulation system that features a thermoplastic insulation layer covered by a heat resistant jacket. Asbestos has been replaced by glass or aramid fiber jackets. Type TA wire provides relatively high ampacities, good temperature performance, and relatively high resistance to abrasion.

**Types.** N/A

### Applications

Used for switchboard applications.

### Sizes

From **#14 AWG up to 4/0 AWG.**

### Temperature Ranges

**Up to 90 degrees Celsius.**

### Voltage Ranges

Circuits **below 600 Volts.**

### Ampacities

Ampacities are determined using standard methods from the *NEC*®.

### Handling and Care

**Receiving and Handling.**
- Examine the protective covering for evidence of damage.
- Do not allow equipment to touch or damage cable surface or protective wrap.

- If a forklift is used, the forks must be long enough to contact both reel flanges.

- If a crane is used, a shaft through the arbor hold or flange cradles must be employed for the lift.

- If an inclined ramp is used for unloading, it must be wide enough to contact both flanges completely. The reel should be stopped at the bottom by using the flanges, not the cable surface.

- Never drop reels on the ground.

**Storage.**

- Reels should be stored on a hard clean surface, not bare earth.

- Store away from open fires or heat sources.

- Cable ends should be resealed as soon as a cable section has been removed.

**Receiving Tests.** Cable should be tested according to the insulation type and the methods outlined in Chapter 4 of this handbook.

## Installation

**Securing and Supporting.** The installation of type TA wire is determined by the particular switchboard application. Cable clamps, cable ties, and lacing twine are all acceptable methods.

**Termination and Splicing.** Type TA wire may be terminated using squeeze-on lugs. Smaller sizes may be terminated directly using methods shown in Figure 6–17.

# THERMOPLASTIC AND FIBROUS OUTER BRAID (TYPE TBS)

## Description

**General.** Thermoplastic and fibrous outer braid is an insulation system that features a thermoplastic insulation layer covered by a fibrous jacket. Type TBS wire provides relatively high ampacities, good temperature performance, and relatively high resistance to abrasion.

**Types.** N/A

## Applications

Used for switchboard applications.

## Sizes

From **#14 AWG up to 4/0 AWG.**

## Temperature Ranges

**Up to 90 degrees Celsius.**

## Voltage Ranges

Circuits **below 600 Volts.**

## Ampacities

Ampacities are determined using standard methods from the *NEC®*.

## Handling and Care

### Receiving and Handling.

- Examine the protective covering for evidence of damage.
- Do not allow equipment to touch or damage cable surface or protective wrap.
- If a forklift is used, the forks must be long enough to contact both reel flanges.
- If a crane is used, a shaft through the arbor hold or flange cradles must be employed for the lift.
- If an inclined ramp is used for unloading, it must be wide enough to contact both flanges completely. The reel should be stopped at the bottom by using the flanges, not the cable surface.
- Never drop reels on the ground.

### Storage.

- Reels should be stored on a surface, not bare earth.
- Store away from open fires or heat sources.
- Cable ends should be resealed as soon as a cable section has been removed.

**Receiving Tests.** Cable should be tested according to the insulation type and the methods outlined in Chapter 4 of this handbook.

## Installation

**Securing and Supporting.** The installation of type TBS wire is determined by the particular switchboard application. Cable clamps, cable ties, and lacing twine are all acceptable methods.

**Termination and Splicing.** Type TBS wire may be terminated using squeeze-on lugs. Smaller sizes may be terminated directly using methods shown in Figure 6–17.

# TRAY CABLE (TYPE TC)

## Description

**General.** Type TC power and control cable is a factory assembly of two or more insulated conductors, with or without bare or covered grounding conductors under a nonmetallic sheath. This type of cable is approved for installation in cable trays, raceways, or where supported by a messenger wire. Figure 6–27 is a typical type TC cable.

**Types.** N/A

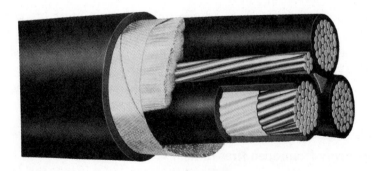

**Figure 6–27   Type TC cable.** *(Courtesy of Rome Cable Corporation)*

## Applications

| Permitted | Not Permitted |
|---|---|
| Power, lighting, and control | Where exposed to physical damage |
| Cable trays | Open cable on brackets or cleats |
| Raceways | Exposed to UV, unless identified |
| Messenger wire supported | Direct buried, unless identified |
| Hazardous locations (318,501,502,504) | |
| Class 1 Circuits (725) | |

### NOTE:

Numbers in parentheses indicate articles in the *NEC*® where additional or explanatory information may be found.

## Sizes

**Copper**

#18 AWG up to 1000 kcmil#

**Aluminum or
Copper Coated Aluminum**

12 AWG up to 1000 kcmil

## Temperature Ranges

The temperature range of type TC cable is determined by the insulation used on the individual conductors. Temperature ranges may be found in the *NEC*®.

## Voltage Ranges

Intended for application at **600 Volts and below**.

## Ampacities

Type TC cable ampacities for conductors larger than #14 AWG are determined using standard methods from the *NEC*®.

## Handling and Care

**Receiving and Handling.**

- Examine the protective covering for evidence of damage.
- Do not allow equipment to touch or damage cable surface or protective wrap.

- If a forklift is used, the forks must be long enough to contact both reel flanges.

- If a crane is used, a shaft through the arbor hold or flange cradles must be employed for the lift.

- If an inclined ramp is used for unloading, it must be wide enough to contact both flanges completely. The reel should be stopped at the bottom by using the flanges, not the cable surface.

- Never drop reels on the ground.

**Storage.**

- Reels should be stored on a hard clean surface, not bare earth.

- Store away from open fires or heat sources.

- Cable ends should be resealed as soon as a cable section has been removed.

**Receiving Tests.** Cable should be tested according to the insulation type and the methods outlined in Chapter 4 of this handbook.

## Installation

**Securing and Supporting.** Tray Cable may only be installed in trays, raceways, or by the method described previously. NEC® Articles 300 and 318 provide installation details.

**Termination and Splicing.** Termination of tray cable is dependent on the type of insulation used in the cable. Splices and terminations are only allowed in the tray using approved boxes or fittings for such an application.

# EXTENDED POLYTETRAFLUOROETHYLENE (TYPE TFE)

## Description

**General.** Type TFE is a special high temperature insulation. It is used for leads connected to apparatus or within raceways connected to apparatus.

**Types.** N/A

## Applications

- Dry locations only.
- Leads within apparatus or in raceways connected to apparatus.
- Open wiring.

## Sizes

From **#14 AWG up to 4/0 AWG.**

## Temperature Ranges

**Up to 250 degrees Celsius.**

## Voltage Ranges

Circuits **below 600 Volts**.

## Ampacities

Ampacities are determined using standard methods from the *NEC*®.

## Handling and Care

**Receiving and Handling.**

- Examine the protective covering for evidence of damage.
- Do not allow equipment to touch or damage cable surface or protective wrap.
- If a forklift is used, the forks must be long enough to contact both reel flanges.
- If a crane is used, a shaft through the arbor hold or flange cradles must be employed for the lift.
- If an inclined ramp is used for unloading, it must be wide enough to contact both flanges completely. The reel should be stopped at the bottom by using the flanges, not the cable surface.
- Never drop reels on the ground.

**Storage.**

- Reels should be stored on a hard clean surface, not bare earth.
- Store away from open fires or heat sources.

- Cable ends should be resealed as soon as a cable section has been removed.

**Receiving Tests.** Cable should be tested according to the insulation type and the methods outlined in Chapter 4 of this handbook.

## Installation

**Securing and Supporting.**  Cable shall be secured and supported as follows:

- Approved staples, straps, hangers or similar fittings, as shown in Figure 6-21.
- When mounted on studs, joists, and rafters, cable shall be installed as shown in Figure 6-22.
- When passed through bored holes, the edge of the hole shall be no less than 1 1/4 inches from the edge of the stud, joist, or rafter.
- In the event that 1 1/4 inches cannot be maintained, as mentioned previously and shown in Figure 6-3, a steel spacer or sleeve must be used to protect the cable.
- Cable installed in accessible attics shall be secured as shown in Figure 6-23.

**Termination and Splicing.**  Figure 6-24 shows the correct methods for termination of type TFE.

## THERMOPLASTIC (TYPE THHN, THHW, THW, THWN, AND TW)

## Description

**General.**  Thermoplastic insulations are modern insulations used in many power and control installations.

**Types.**  Heat resistant thermoplastic cable is available in the following types.

| | |
|---|---|
| **THHN** | Flame retardant, heat resistant, with nylon jacket. |
| **THHW** | Flame retardant, moisture and heat resistant. |
| **THW** | Flame retardant, moisture and heat resistant. |
| **THWN** | Flame retardant, moisture and heat resistant, with nylon jacket. |
| **TW** | Flame retardant, moisture resistant. |

## Applications

Used for all general wiring requirements for low and medium voltage applications.

## Sizes

Types THW, TW: From **#14 AWG up to 2000 kcmil.**
Types THHN, THHW, THWN: From **#14 AWG up to 1000 kcmil.**

## Temperature Ranges

**Up to 90 degrees Celsius.**

## Voltage Ranges

Circuits **below 600 Volts up to 35,000 Volts.**

## Ampacities

Ampacities are determined using standard methods from the *NEC®*.

## Handling and Care

**Receiving and Handling.**
- Examine the protective covering for evidence of damage.
- Do not allow equipment to touch or damage cable surface or protective wrap.
- If a forklift is used, the forks must be long enough to contact both reel flanges.
- If a crane is used, a shaft through the arbor hold or flange cradles must be employed for the lift.
- If an inclined ramp is used for unloading, it must be wide enough to contact both flanges completely. The reel should be stopped at the bottom by using the flanges, not the cable surface.
- Never drop reels on the ground.

**Storage.**
- Reels should be stored on a surface, not bare earth.
- Store away from open fires or heat sources.

- Cable ends should be resealed as soon as a cable section has been removed.

**Receiving Tests.** Cable should be tested according to the insulation type and the methods outlined in Chapter 4 of this handbook.

## Installation

**Securing and Supporting.** Low-voltage cable shall be secured and supported as follows.

- Approved staples, straps, hangers or similar fittings, as shown in Figure 6-21.
- When mounted on studs, joists, and rafters, cable shall be installed as shown in Figure 6-22.
- When passed through bored holes, the edge of the hole shall be no less than 1 1/4 inches from the edge of the stud, joist, or rafter.
- In the event that 1 1/4 inches cannot be maintained, as mentioned previously and in Figure 6-3, a steel spacer or sleeve must be used to protect the cable.
- Cable installed in accessible attics shall be secured as shown in Figure 6-23.

High-voltage cable should be installed as described previously for type MV cable.

**Termination and Splicing.** Terminations depend upon application voltages. See the appropriate sections for complete information.

## UNDERGROUND FEEDER AND BRANCH CIRCUIT CABLE (TYPE UF)

### Description

**General.** Type UF is a fabricated assembly of one or more individually insulated conductors. Insulation used must be of a moisture resistant type.

**Types.** N/A

## Applications

### Permitted
Underground direct burial
Interior wiring in all locations

### Not Permitted
Service entrance cable
Commercial garages
Theaters
Motion picture studios
Storage battery rooms
Hoistways
Hazardous locations
Embedded in poured concrete (424)
Exposed to sunlight, unless identified

### NOTE:

Numbers in parentheses indicate articles in the *NEC*® where additional or explanatory information may be found.

## Sizes

### Copper

#14 AWG up to 4/0 AWG

### Aluminum or Copper Coated Aluminum

#12 AWG up to 4/0 AWG

## Temperature Ranges

60 degrees Celsius.

## Voltage Ranges

600 Volts and lower.

## Ampacities

Type UF cable ampacities are determined using the 60 degree Celsius conductor entries from the current edition of the *NEC*®.

## Handling and Care

### Receiving and Handling.

- Examine the protective covering for evidence of damage.
- Do not allow equipment to touch or damage cable surface or protective wrap.
- If a forklift is used, the forks must be long enough to contact both reel flanges.
- If a crane is used, a shaft through the arbor hold or flange cradles must be employed for the lift.
- If an inclined ramp is used for unloading, it must be wide enough to contact both flanges completely. The reel should be stopped at the bottom by using the flanges, not the cable surface.
- Never drop reels on the ground.

### Storage.

- Unjacketed, metal-clad cable should be stored indoors to avoid corrosion.
- Reels should be stored on a hard clean surface, not bare earth.
- Store away from open fires or heat sources.
- Cable ends should be resealed as soon as a cable section has been removed.

**Receiving Tests.** Cable should be tested according to the insulation type and the methods outlined in Chapter 4 of this handbook.

## Installation

**Securing and Supporting.** When used for interior wiring, type UF cable shall be secured and supported as follows.

- Approved staples, straps, hangers or similar fittings, as shown in Figure 6-10.
- In cable trays where approved for that application.
- Direct buried as per Figure 6-11.
- Type UF cable may be used as overhead or underground service entrance cable.

**Termination and Splicing.** Type UF cable shall be terminated and spliced according to the low voltage methods presented in Chapter 3 of this handbook. Underground splices shall be made using methods which render the cable impervious to moisture.

## VARNISHED CAMBRIC (TYPE V)

### Description

**General.** Varnished cambric is an insulated cable composed of conductors wrapped with cotton tape which has been coated with insulating varnish. The layers of tape are separated with a layer of nonhardening mineral compound. The mineral compound serves as a lubricant.

If the cable is to be used in a wet location, it *must* be covered with a moisture-proof jacket, such as lead or thermoplastic.

Varnished cambric is an older insulation whose electrical characteristics fall somewhere between paper and thermoplastic.

**Types.**  N/A

### Applications

Used for all general wiring requirements for low- and medium-voltage applications in dry locations only.

### Sizes

From **#14 AWG up to 2000 kcmil.** Sizes smaller than #6 AWG are by special permission only.

### Temperature Ranges

**Up to 85 degrees Celsius.**

### Voltage Ranges

Circuits **below 600 Volts up to 35,000 Volts.**

### Ampacities

Ampacities are determined using standard methods from the NEC®.

## Handling and Care

**Receiving and Handling.**

- Examine the protective covering for evidence of damage.
- Do not allow equipment to touch or damage cable surface or protective wrap.
- If a forklift is used, the forks must be long enough to contact both reel flanges.
- If a crane is used, a shaft through the arbor hold or flange cradles must be employed for the lift.
- If an inclined ramp is used for unloading, it must be wide enough to contact both flanges completely. The reel should be stopped at the bottom by using the flanges, not the cable surface.
- Never drop reels on the ground.

**Storage.**

- Reels should be stored on a hard clean surface, not bare earth.
- Store away from open fires or heat sources.
- Cable ends should be resealed as soon as a cable section has been removed.
- Type V cable should never be stored in a moist or wet area.

**Receiving Tests.** Cable should be tested according to the insulation type and the methods outlined in Chapter 4 of this handbook.

## Installation

**Securing and Supporting.** Low-voltage cable shall be secured and supported as follows.

- Approved staples, straps, hangers or similar fittings, as shown in Figure 6-21.
- When mounted on studs, joists, and rafters, cable shall be installed as shown in Figure 6-22.
- When passed through bored holes, the edge of the hole shall be no less than 1 1/4 inches from the edge of the stud, joist, or rafter.

- In the event that 1 1/4 inches cannot be maintained, as mentioned previously and in Figure 6–3, a steel spacer or sleeve must be used to protect the cable.

- Cable installed in accessible attics shall be secured as shown in Figure 6–23.

High-voltage cable should be installed as described previously for type MV cable.

**Termination and Splicing.** Terminations depend upon application voltages. See the appropriate sections for complete information.


# CROSS-LINKED POLYMER (TYPES XHHW AND XHHW-2)

## Description

**General.** An insulation made of a cross-linked polymer, such as polyethylene. These types of thermosetting insulations have superior temperature characteristics.

**Types.** Types XHHW and XHHW-2 are both cross-linked polymer types. Type XHHW-2 is a newer listing, and is allowed in wet or dry installations to 90 degrees Celsius.

## Applications

Used for all general wiring requirements for low- and medium-voltage applications.

## Sizes

From **#14 AWG up to 2000 kcmil.**

## Temperature Ranges

**Type XHHW:**     Dry locations **up to 90 degrees Celsius.**
                                Wet locations **up to 75 degrees Celsius.**
**Type XHHW-2:**  Wet or dry locations **up to 90 degrees Celsius.**

## Voltage Ranges

Circuits **below 600 Volts up to 35,000 Volts.**

## Ampacities

Ampacities are determined using standard methods from the NEC®.

## Handling and Care

**Receiving and Handling.**

- Examine the protective covering for evidence of damage.
- Do not allow equipment to touch or damage cable surface or protective wrap.
- If a forklift is used, the forks must be long enough to contact both reel flanges.
- If a crane is used, a shaft through the arbor hold or flange cradles must be employed for the lift.
- If an inclined ramp is used for unloading, it must be wide enough to contact both flanges completely. The reel should be stopped at the bottom by using the flanges, not the cable surface.
- Never drop reels on the ground.

**Storage.**

- Reels should be stored on a hard clean surface, not bare earth.
- Store away from open fires or heat sources
- Cable ends should be resealed as soon as a cable section has been removed.

**Receiving Tests.** Cable should be tested according to the insulation type and the methods outlined in Chapter 4 of this handbook.

## Installation

**Securing and Supporting.** Low-voltage cable shall be secured and supported as follows.

- Approved staples, straps, hangers or similar fittings, as shown in Figure 6-21.

- When mounted on studs, joists, and rafters, cable shall be installed as shown in Figure 6–22.
- When passed through bored holes, the edge of the hole shall be no less than 1 1/4 inches from the edge of the stud, joist, or rafter.
- In the event that 1 1/4 inches cannot be maintained, as mentioned previously and shown in Figure 6–3, a steel spacer or sleeve must be used to protect the cable.
- Cable installed in accessible attics shall be secured as shown in Figure 6–23.

High-voltage cable should be installed as described previously for type MV cable.

**Termination and Splicing.** Terminations depend upon application voltages. See the appropriate sections for complete information.

## MODIFIED ETHYLENE TETRAFLUOROETHYLENE (TYPES Z AND ZW)

### Description

**General.** Types Z and ZW are insulation systems which have excellent temperature and radiation characteristics. Figure 6–28 shows a typical type Z cable.

**Types.** Modified Ethylene Tetrafluoroethylene is available in the following types:

| | |
|---|---|
| Z | Dry and damp locations. |
| | Dry locations; special applications. |
| ZW | Wet locations. |
| | Dry and damp locations. |
| | Dry locations; special applications. |

THE OKONITE CO. PLT 6 1/C 10 AWG CU OKOZEL (EFTE) 600V UL TYPE Z

**Figure 6–28   Type Z wire.** *(Courtesy of The Okonite Company —Power, Control, and Instrumentation Cables)*

## Applications

Type Z cables are used in areas such as nuclear power plants and fossil fuel plants where ruggedness, high temperature, and radiation resistance is important.

## Sizes

**Type Z:**   From #14 AWG up to 4/0 AWG.
**Type ZW:** From #14 AWG up to #2 AWG.

## Temperature Ranges

**Type Z:**   **Up to 90 degrees Celsius** for dry and damp locations.
             **Up to 150 degrees Celsius** for dry locations and special applications.
**Type ZW: Up to 75 degrees Celsius** for wet locations.
             **Up to 90 degrees Celsius** for dry and damp locations.
             **Up to 150 degrees Celsius** for dry locations and special applications.

## Voltage Ranges

Circuits **below 600 Volts.**

## Ampacities

Ampacities are determined using standard methods from the *NEC®*.

## Handling and Care

**Receiving and Handling.**

- Examine the protective covering for evidence of damage.
- Do not allow equipment to touch or damage cable surface or protective wrap.
- If a forklift is used, the forks must be long enough to contact both reel flanges.
- If a crane is used, a shaft through the arbor hold or flange cradles must be employed for the lift.
- If an inclined ramp is used for unloading, it must be wide enough to contact both flanges completely. The reel should be

stopped at the bottom by using the flanges, not the cable surface.

- Never drop reels on the ground.

**Storage.**

- Reels should be stored on a hard clean surface, not bare earth.
- Store away from open fires or heat sources.
- Cable ends should be resealed as soon as a cable section has been removed.

**Receiving Tests.** Cable should be tested according to the insulation type and the methods outlined in Chapter 4 of this handbook.

## Installation

**Securing and Supporting.** Low-voltage cable shall be secured and supported as follows.

- Approved staples, straps, hangers or similar fittings, as shown in Figure 6-21.
- When mounted on studs, joists, and rafters, cable shall be installed as shown in Figure 6-22.
- When passed through bored holes, the edge of the hole shall be no less than 1 1/4 inches from the edge of the stud, joist, or rafter.
- In the event that 1 1/4 inches cannot be maintained, as mentioned previously and shown in Figure 6-3, a steel spacer or sleeve must be used to protect the cable.
- Cable installed in accessible attics shall be secured as shown in Figure 6-23.

High-voltage cable should be installed as described previously for type MV cable.

**Termination and Splicing.** Termination of these types is standard to low voltage applications. See Chapter 3 of this handbook.

# Index

Note: Page references in **bold** type reference non-text material.